Mark Anthony Benvenuto, Heinz Plaumann
Industrial Catalysis

Also of Interest

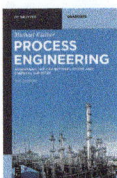

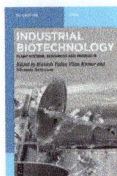

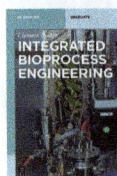

Mark Anthony Benvenuto, Heinz Plaumann

Industrial Catalysis

—

DE GRUYTER

Authors
Prof. Dr. Mark Anthony Benvenuto
Department of Chemistry and Biochemistry
University of Detroit Mercy
4001 W. McNichols Rd.
Detroit, MI 48221-3038
USA
benvenma@udmercy.edu

Heinz Plaumann, Ph.D.
CEO and Co-Founder Quantum Qik Careers
734-747-4770
www.quantumqik.com
heinz.quantumqik@gmail.com

ISBN 978-3-11-054284-4
e-ISBN (PDF) 978-3-11-054286-8
e-ISBN (EPUB) 978-3-11-054294-3

Library of Congress Control Number: 2021936050

Bibliographic information published by the Deutsche Nationalbibliothek
The Deutsche Nationalbibliothek lists this publication in the Deutsche Nationalbibliografie;
detailed bibliographic data are available on the Internet at http://dnb.dnb.de.

© 2021 Walter de Gruyter GmbH, Berlin/Boston
Cover image: 1971yes/iStock/Getty Images Plus
Typesetting: Integra Software Services Pvt. Ltd.
Printing and binding: CPI books GmbH, Leck

www.degruyter.com

Contents

Chapter 1
Introduction

1.1 Catalysis

The acceleration of a chemical reaction by some chemical that is itself not consumed in the process has been the working definition of catalysis for over a century.
It is probably not a coincidence that the development of catalysts roughly parallels
or follows the expansion of the Industrial Revolution. This revolution brought with
it not only the larger scale production of commodities than had ever been accomplished before, it also coupled the production of commodities to making money and
the creation of wealth. Prior to this, civilizations had flourished using what can be
called cottage industries to make virtually all end products. For example, smelting
and working metals for the production of coins, tools, armor, and weaponry was a
large enough enterprise in the Roman Empire that its aerial pollution and effluents
appear to have in some manner polluted a large swath of Europe, and left a trace in
even the Greenland ice cap. But this was the result of more individuals working at
trades – again, cottage industries – and not the mechanization of any processes. During the Industrial Revolution, trade over larger areas than had ever been seen before
meant that larger amounts of materials, commodities, and end products were needed.
That in turn meant that less expensive methods of making materials were required.
As the nineteenth century turned to the twentieth, this in turn began to mean that the
use of a catalyst to produce a product was economically advantageous.

As any chemical field blossoms, there are eventually names associated with
some portion of it, often with a specific advance or a specific reaction. In the case of
catalysis, one scientist who can be considered the father of the field is Vladimir Nikolayevich Ipatieff. Born in Imperial Russia in 1867, Ipatieff became important to
the Czar's war effort in the First World War, realized that the political environment
had become extremely dangerous by the end of the 1920s, and ultimately emigrated
to the west in 1930. Toward the beginning of what became a long and fruitful career,
he noticed that some reactions proceeded much more quickly in steel containers
than they did in glass – the first mention of a serendipitous catalysis. Later in his
career, working at Universal Oil Products (UOP), Ipatieff led teams that produced
numerous catalysts, some of which are still in use today. While Professor Ipatieff's
name is unfortunately not well known or remembered among the general population, or even chemists, today, he did leave a lasting legacy. His work with UOP was
fruitful enough that the company made a significant endowment to the American
Chemical Society, which was used to present the Ipatieff Prize every 3 years. From
its first presentation in 1947, the prize has become an impressive list of some of the
greatest researchers in the field of catalysis.

https://doi.org/10.1515/9783110542868-001

1.2 Reaction chemistry

Curiously, despite the growing use of catalysts over the course of time, and despite the standardization of what we now think of as basic reaction chemistry, there has never been a standardized, uniform way to represent a catalyst in a written chemical reaction. Figure 1.1 shows the two means by which a catalyst is normally indicated in a reaction, above or below the reaction itself. Note that some professionals will debate whether or not a subscripted indicator like "(cat)" is necessary, while others will insist that one of the two representations is the only official or correct one. But the fact remains that a catalyst is often represented in one of these two ways, but may also be represented by other means as well. It is also noteworthy that the state or conditions of a catalyst cannot be easily represented in a reaction. Again turning to Figure 1.1, it is not known whether the iron catalyst is a powder, a foil, or a material on some support.

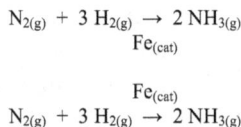

$$N_{2(g)} + 3\,H_{2(g)} \rightarrow 2\,NH_{3(g)}$$
$$Fe_{(cat)}$$

$$Fe_{(cat)}$$
$$N_{2(g)} + 3\,H_{2(g)} \rightarrow 2\,NH_{3(g)}$$

Figure 1.1: Representations of a catalyst in a reaction.

As well, almost universally, while the presence of a catalyst is shown in written reactions, the conditions under which it operates are not. Does the catalyst reduce the working temperature of a reaction, for example? While that is a generally favorable reason to use a catalyst, temperatures are not routinely mentioned in written representations of reactions. Does the catalyst only function under high pressure and high temperature (albeit, lower than the un-catalyzed reaction)? Again, this is not always shown in a written representation of a reaction, although it can be placed above or below a reaction arrow as well. The means by which a catalyst and reaction conditions are represented continue to have no standardized type of notation.

1.3 Catalyst production

The production of catalysts, as opposed to their uses, is the main focus of this book. We must arrange it by the chemical product, or by the chemical process, being discussed, but the main idea and focus is always, how is the catalyst itself produced? The answer to that question is not only a matter for and of corporate research but is the main thrust of numerous national and international societies [1–9].

Table 1.1 is a non-exhaustive list of major catalyst producing companies. Note that some of the names are rather common to chemists and chemical engineers,

Table 1.1: Major catalyst producers.

Company name	Catalyst types	Other products	Website
Albermarle	Petrochemical, oil upgrading	Specialty chemicals, lithium, bromine	www.albermarle.com
Axens	Related to petroleum refining	Oil, natural gas, biomass-based fuels	www.axens.net
BASF	"Environmental and process catalysts" Emission control	Wide variety of chemical commodities	www.basf.com
Criterion	Refinery operations	Petroleum	www.shell.com
ExxonMobil	Gas and refining	Petroleum	www.corporate.exxon mobil.com
Grace Davison	Petroleum refining		www.grace.com
Haldor Topsoe	For oil and gas refining	Supplies catalysts to customers	www.topsoe.com
Johnson Matthey	Powders	Metals	matthey.com
Sűd Chemie	Base metal catalysts	Power technology	clariant.com
UOP	Catalysts for petroleum products		uop.com

From Refs. [10–20].

while others are less well known. For example, ExxoMobil is known to the public as a gas and petroleum company. Yet it produces a far greater number of products than just motor gasoline. Similarly, Johnson Matthey is known by many as a metal distributor but has a catalyst division.

Some companies, such as Albermarle, Axens, Sud Chemie, and UOP, are known to scientists and engineers but generally remain unknown to the general public. But all produce catalysts, sometimes for many applications, and often for other companies.

1.4 Catalyst fate

One of the most basic aspects of chemistry is the law of conservation of mass. Matter is converted through chemical reactions, but neither created nor destroyed. Since a catalyst promotes a chemical reaction but does not become part of the product, we sometimes do not consider it in terms of this law. Yet in any reaction the catalyst simply must be *somewhere* when the reaction is completed. The material that makes up the catalyst cannot simply disappear.

Catalysts that exist as a solid surface and that interact while being in a different phase from the reactants – heterogeneous catalysts – often stay in place no matter how much reactant comes in contact with them. They may lose effectiveness if sites are poisoned because of prolonged use and the presence of even tiny amounts of impurities but do not generally move. Homogeneous catalysts, however, or solid catalysts that exist as powders for maximum surface area, can migrate with the reactants as those reactants become products, and sometimes can become embedded in the product. The possibility for this is highest when a liquid or solvated reactant in a flow-through system becomes a solid product.

The recovery of catalysts becomes important when the value of a particular catalyst material is high, and the availability of it is low, or can be disrupted through some problem in its supply chain from source to user. Platinum serves as an excellent example of both of these considerations. It is an expensive transition metal, and its sourcing and production are tracked by governments to ensure its availability [21]. Recovery of platinum from any process in which it is used can be essentially a re-refining of it. The same holds true for several other scarce materials that show a catalytic activity.

References

[1]   North American Catalysis Society (NACS). Website. (Accessed 22 February 2020, as: www.nacatsoc.org).
[2]   European Federation of Catalysis Societies (EFCATS). Website. (Accessed 22 February 2020, as: efcats.org).
[3]   Asia-Pacific Association of Catalysis Societies (APACS). Website. (Accessed 22 February 2020, as: www.apacs.dicp.ac.cn).
[4]   Iberoamerican Federation of Catalysis Societies (FISOCAT). Website. (Accessed 22 February 2020, as: www.fiq.unl.edu.ar/fisocat).
[5]   Catalysis Society of Japan. Website. (Accessed 22 February 2020, as: www.shokubai.org).
[6]   Nordic Catalysis Society. Website. (Accessed 22 February 2020, as: www.nordic-catalysis.org).
[7]   The China Catalysis Society. Website. (Accessed 22 February 2020, as: www.catalysis.org.cn).
[8]   International Zeolite Association (IZA). Website. (Accessed 22 February 2020, as: http://www.iza-online.org).
[9]   International Association of Catalysis Societies (IACS). Website. (Accessed 22 February 2020, as: iacs-catalysis.org).
[10] Middle East Oil and Gas.com. Website. (Accessed 9 October 2018, as: https://www.oilandgasmiddleeast.com/article-9496-top-10-catalysts-companies).
[11] Albermarle. Website. (Accessed 27 March 2019, as: www.albermarle.com/business/refining-solutions).
[12] Axens. Website. (Accessed 27 March 2019, as: mobile.axens.net).
[13] BASF. Website. (Accessed 9 October 2018, as: https://catalysts.basf.com/).

[14] Criterion. Website. (Accessed 27 March 2019, as: www.criterioncatalysts.com).
[15] ExxonMobil Chemical. Website. (Accessed 27 March 2019, as: www.exxonmobilchemical.com).
[16] Grace Davison. Website. (Accessed 27 March 2019, as: grace.com/en-us).
[17] Haldor Topsoe. Website. (Accessed 27 March 2019, as: www.topsoe.com/products/catalysts).
[18] Johnson Matthey. Website. (Accessed 27 March 2019, as: matthey.com/products).
[19] Süd-Chemie. Website. (Accessed 27 March 2019, as: www.clariant.com/en/Company).
[20] UOP. Website. (Accessed 27 March 2019, as: www.uop.com/products).
[21] Mineral Commodity Summaries. Downloadable as: https://www.usgs.gov/centers/nmic/mineral-commodity-summaries

Chapter 2
Homogeneous catalysis

2.1 Introduction

All of the catalysts used in any industrial-scale process fall into one of two very broad categories: heterogeneous or homogeneous. We will start our discussion with the homogeneous class. The term "homogeneous" means that the catalyst is in the same phase as the material with which it interacts. Thus, most of such catalysts will be soluble materials for use in the liquid phase. The material that reacts is either a neat liquid, or more often, a solute that is soluble in some liquid solvent.

It is almost impossible to discuss catalysts and catalysis without using some example, and so we will use the wide range of materials that fall under the heading of Ziegler–Natta catalysts for some examples. Named after the two pioneers of this class of catalysts – Karl Ziegler and Gullio Natta – soluble Ziegler–Natta catalysts are often a metallocene molecule, as shown in Figure 2.1, used with some mixture of an alumino-organic compound, such as methylaluminoxane (used so often that it is abbreviated MAO). The metallocene routinely incorporates an element from the early transition metals, such as titanium, zirconium, or hafnium, although others have been tried.

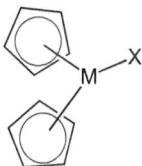

Figure 2.1: Example of a metallocene catalyst.

Note that the choice of metal can be different for different reactions, and the organic portion of the metallocene can be substituted as well, meaning that there are numerous possibilities when substituted cyclopentadienyl rings are used. Additionally, various "pieces" of the molecule can be connected directly to the metal. The reason for this wide variety of possibilities in a catalyst is that they are all simply adjusted until the best working catalyst for a specific reaction is found – the one with the highest turnover number. There is no overall theory of catalyst synthesis.

The term "turnover number" is a measure of how many times a reaction comes to completion using a catalyst. When a reaction is 100% complete, the turnover number is 1. Perhaps obviously, high turnover numbers (in millions, if they can be measured) are desirable.

https://doi.org/10.1515/9783110542868-002

Also, although Figure 2.1 shows two organic portions, the cyclopentadienyl rings, at such an angle that the rings seem to be lying on the plane of the page, they form what is popularly called a sandwich with the metal atom or ion in the center. Figure 2.2 illustrates this.

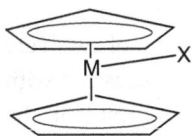

Figure 2.2: Metallocene molecule shown as a sandwich compound.

The two rings can be either eclipsed or staggered in relation to each other, and often interconvert configurations when undergoing a catalytic activity. A number of possibilities exist for what "X" can be in such a system.

2.2 Advantages of a homogeneous system

2.2.1 Reaction chemistry

The main advantage of any homogeneous catalyst is that as many atoms of the catalyst as possible are able to interact with the reactants. Consider, for a moment, a heterogeneous catalyst that exists as a foil. No matter how thin the foil is, any atom below the surface of the foil can be considered a "dead" atom, unable to affect any catalysis. But with a homogeneous catalyst, at least in theory, every catalyst atom is available to promote the reaction.

This is a somewhat idealized version of a homogeneous catalyst, since materials have different solubility levels in different solvents. A catalyst with a high turnover number may not be particularly good in one solvent, thus limiting its overall results. Despite this, however, the main advantage of any homogeneous catalyst is its ability to meet and react with the reagents.

2.2.2 Catalyst production

The production of a catalyst, as opposed to its use, can be a proprietary one. However, in very broad terms, the metallocene catalyst example used in Figure 2.1 can be broken down into a few general steps. They are:
1. Refining and production of the metal salt
2. Synthesis of the cyclopentadienyl rings (often called Cp rings)
3. Production or assembly of the metallocene

Several companies, as mentioned in Chapter 1, have divisions that specialize in the production and manufacture of catalysts. But the three steps that were just mentioned can be further specified, as follows:

1. The desired metal is refined and reduced from an ore. The refined metal is then oxidized to a specific salt form, for example: $M + X_2 \rightarrow MX_2$, where M is the metal and X is a nonmetallic element.
2. The cyclopentadiene portion is extracted either from coal tar, or can be manufactured through what is called the steam cracking of the naphtha fraction of crude oil. Since it is obtained as a dimer, it must be heated to 150–200 °C, then distilled and used.
3. The Cp ring exists as a neutral C_5H_6 organic molecule and is deprotonated, forming Cp^- (formula $C_5H_5^-$), then added to the metal salt in a two-to-one ratio, forming the metallocene.

Even this level of explanation can be considered very general, as many different substituents can be part of the Cp rings, and a wide variety of metal salts can be used to incorporate the metal atom into the structure.

2.3 Catalyst fate

The use of rare and valuable materials in many catalytic processes is one important reason that they are tracked, and their fate is examined. We have seen that a major advantage of a homogeneous catalysts system is that any molecule containing the precious metal is part of an active catalytic site. In simple terms: all the catalyst is used.

If there is a negative aspect to homogeneous catalytic systems, it is that the catalyst is not stationary or in some way anchored to one site. The catalyst is subject to movement and flow as the product is made. Thus, over the course of time, the catalyst can migrate and can become part of the product. For this reason, batches of some commodity chemicals are assayed, to determine if some tiny amount of residual catalyst exists within them.

Chapter 3
Heterogeneous catalysis

3.1 Introduction

To those interested in the study of catalysis and reactions that can be improved based on some catalytic process, there are advantages and disadvantages to both homogeneous and heterogeneous catalysts. A heterogeneous catalyst is defined as one that is not in the same phase as the reaction it catalyzes. In most cases, this means the catalyst is a solid, and the reaction chemistry which it catalyzes is in a gaseous or liquid state. But whatever be the state of the catalyst and the reactants, the function of the catalyst remains to lower the activation energy of the reaction, which routinely translates to an economic benefit when producing a product.

The entire broad field of catalysis has its origins with heterogeneous catalysts, and the work of Vladimir Nikolayevich Ipatieff, mentioned briefly in Chapter 1, who recognized that the wall of an iron container promoted chemical reactions, whereas the same reactions run in quartz or glass containers did not proceed nearly as quickly. Ipatieff originally worked for Czar during the First World War, in the production of chemical munitions, then later defected from the Soviet Union to the United States in 1930, again as mentioned in Chapter 1, and continued and expanded his work until his death in 1952. Numerous large-scale reactions today continue to use catalysts originally discovered by Ipatieff during his time working at UOP.

3.2 Reaction chemistry

Rules for the representation of a chemical reaction have evolved over the course of decades, and as mentioned the use of a catalyst in a reaction is often represented above or below the reaction arrow, as shown in Figure 3.1. Since the catalyst never becomes part of the product, it is not shown alongside the reactants or the products.

$$A + B \xrightarrow{\text{catalyst}} C + D$$

$$A + B \xrightarrow[\text{catalyst}]{} C + D$$

Figure 3.1: Representation of a catalyst in a chemical reaction.

In whatever manner the catalyst is represented, it functions by inducing a reaction to occur on its surface, by lowering the energy of activation for the reaction, and thus the

https://doi.org/10.1515/9783110542868-003

surface area for any heterogeneous catalytic material becomes extremely important. There must be some adsorption of reactants on the catalyst surface, followed by bond breaking, new bond formation, and finally release of the product from the catalyst. Figure 3.2 shows this in a very broad way.

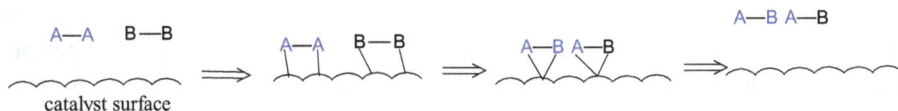

Figure 3.2: Steps in a heterogeneously catalyzed reaction.

The Haber process for the formation of ammonia from its elements is a now-classic example of a heterogeneous catalyst being needed in a reaction. Figure 3.3 shows the basic reaction chemistry.

$$N_{2(g)} + 3\,H_{2(g)} \longleftrightarrow 2\,NH_{3(g)}$$

Figure 3.3: Ammonia production.

The catalyst, which can be several different metals, but is often iron or iron-based, exists as tiny spheres and is brought into contact with the reactant gases at high temperature and pressure. Even at elevated temperatures and pressures, the reaction does not proceed to completion in a single contact to single pass, and the gases are recycled and re-introduced to the reaction chamber. It is therefore important that the catalyst be present and able to make contact with the reactants multiple times.

3.3 Catalyst production

The production and manufacture of heterogeneous catalysts can often be a proprietary matter, especially if the process is one in which several large corporations compete with each other for market share of the same product. But broadly, the following are important aspects of the production of heterogeneous catalysts:
1. Maximizing the surface area of the catalyst
2. Utilizing the most cost-effective materials for the catalyst
3. Ensuring as much of the catalyst can be recovered as possible

The comment about catalysts being made in a cost-effective manner may seem trivial, but there is a great deal of importance attached to it. Since many catalysts are rare elements, their production, refining, and use are tracked by national organizations, such at the U.S. Geological Survey in their annual Mineral Commodity Summaries [1], as well as the U.S. Department of Energy's Critical Materials Strategy [2]. Additionally, the production of refined metals for use as catalysts (and for other applications) is monitored for its environmental footprint [3]. This has become more important and transparent in the past few decades, as elements like cerium and other heavy metals are increasingly being used in catalytic converters, as an example.

3.4 Catalyst fate

The fate of heterogeneous catalysts is often that they are recovered. While a catalyst may not function at maximum efficiency after it has been used extensively, and recovered, the presence of a precious metal or rare element usually dictates that the catalyst be regenerated. Regeneration can take different forms for different materials, but perhaps obviously, must culminate in the catalyst again being functional for its originally intended process.

References

[1]   U.S. Geological Survey. Mineral Commodity Summaries, downloadable.
[2]   U.S. Department of Energy. Critical Materials Strategy, downloadable.
[3]   The dystopian lake filled by the world's tech lust. bbc.com. Website. (Accessed 6 April 2020, as: https://getpocket.com/explore/item/the-dystopian-lake-filled-by-the-world-s-tech-lust?utm_source=pocket-newtab).

Chapter 4
Acrylics

4.1 Introduction

The term "acrylics" encompasses a series of polymers all of which have in common some acrylate monomer. They are also referred to as polyacrylates or acrylate polymers. What they have in common beyond their carbon–carbon atom polymer backbone is a pendant side chain that is an ester in all cases where the R group is not a proton. They are produced in large enough quantities that they are considered commodity chemicals that are of interest to several national and international trade associations [1–5]. The Lewis structure of the repeating unit is shown in Figure 4.1.

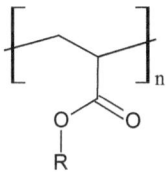

Figure 4.1: Acrylics repeating unit.

The nature of the R group in the polymer has a profound effect on the physical properties of any resulting polymeric material. This is why acrylics are often referred to in the plural, because changing the side chain changes the actual polymer that is produced.

4.2 Reaction chemistry

Figure 4.2 shows the basic reaction that produces an acrylic from the monomer, called an acryloyl monomer. As mentioned, if the R in the acryloyl group is a hydrogen atom, the group is then acrylic acid.

Figure 4.2: Polymerization of the acryloyl monomer.

https://doi.org/10.1515/9783110542868-004

It is difficult to represent the role of any catalyst in a reaction such as that shown in Figure 4.2. But we see at this point, and will see in several later chapters of this book, that the catalytic activity is related to the olefinic or double bond that forms the backbone or main chain of the polymer. The polymerization of this monomer can be affected with a wide variety of catalysts, and interest remains high in determining if more effective catalysts can still be produced [6]. The R group has some influence upon which catalysts work best for specific polymerizations. In the example of H as the R group, polyacrylic acid can be produced from the monomer using potassium persulfate and iron (II) sulfate pentahydrate, the former as an initiator, and the latter as a means of regulating chain length.

4.3 Catalyst production

4.3.1 Potassium persulfate

$K_2S_2O_8$ can be produced as shown in Figure 4.3. The reaction is an electrolysis of potassium hydrogen sulfate (aka potassium bisulfate), and requires an acidic solution, of sulfuric acid.

$$2\ KHSO_{4(aq)} \rightarrow H_{2(g)} + K_2S_2O_8$$

Figure 4.3: Potassium persulfate production.

The scission of the O–O bond in the persulfate produces two sulfate radicals which initiate the polymerization by attack at the double bond.

4.3.2 Iron (II) sulfate pentahydrate

The mineral ferrohexahydrite does exist naturally, but production in acidic solution is the means by which enough is produced for all industrial- and large-scale uses. While more than one means exists to produce $FeSO_4$ of any type of hydrate, perhaps the largest one is as a by-product in the production of steel. Figure 4.4 shows the basic chemistry.

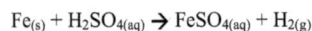

$$Fe_{(s)} + H_2SO_{4(aq)} \rightarrow FeSO_{4(aq)} + H_{2(g)}$$

Figure 4.4: Iron (II) sulfate production.

We will see in a later chapter how sulfuric acid is produced. It also requires a catalyst in its synthetic pathway.

4.4 Catalyst fate

These catalysts are currently inexpensive enough that recovery of them is not common. When disposal costs become a concern, the materials can be recovered, although their regeneration for reuse as catalysts does not always occur.

References

[1]   U.S. Plastics. Website. (Accessed 4 October 2020, as: www.usplastic.com).
[2]   Canadian Plastics. Website. (Accessed 4 October 2020, as: www.canplastics.com).
[3]   PlasticsEurope. Website. (Accessed 4 October 2020, as: www.plasticseurope.org).
[4]   Plastics – SA. Website. (Accessed 4 October 2020, as: www.plasticsinfo.co.za).
[5]   Plastics Stewardship Australia. Website. (Accessed 4 October 2020, as: www.plastics. org.au).
[6]   American Chemical Society, Catalyst Division. Website. (Accessed 4 October 2020, as: www.acs-catalysis.org/about).

Chapter 5
Adipic acid

5.1 Introduction

The production of adipic acid is a large-scale industrial process that results in over 2.5 billion kilograms, or 2.5 million metric tons, of product per year. The vast majority of it is further consumed in the production of nylon, often nylon-6,6. Curiously, adipic acid does have a food additive number E355, even though this is a minor use of the material. Figure 5.1 shows the Lewis structure of adipic acid.

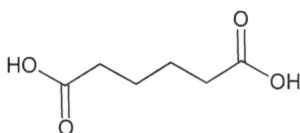

Figure 5.1: Adipic acid.

There is more than one method for the production of adipic acid from some materials that ultimately come from distillates of crude oil. A significant amount is produced from cyclohexane, which in turn is hydrogenated from benzene. This is shown in simplified form in Figure 5.2. This hydrogenation step requires Raney nickel as a catalyst.

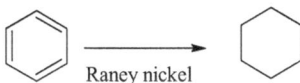

Figure 5.2: Benzene to cyclohexane.

Benzene is made into cyclohexane at 370–800 psi despite the use of the catalyst. The Raney Ni catalyst is used at elevated temperature as well, approximately 200 °C.

5.2 Catalyst production

Raney nickel, sometimes still called spongy nickel, is actually a nickel–aluminum alloy, first developed in 1927, by Murray Raney [1]. It is a heterogeneous catalyst but is not purely nickel with a large surface area. Rather, it is an alloy of nickel and aluminum. Its production steps are as follows, since there is really no stoichiometric mixings that could be easily represented with a series of reactions:

https://doi.org/10.1515/9783110542868-005

1. Ni and Al are mixed in the melt, usually in a 50:50 mixture by mass.
2. The molten elements are quenched in water to create a grainy material with high surface area, often still termed "shot."
3. Small amounts of Zn or Cr may be added to enhance activity.
4. The material is treated with NaOH, of concentration up to 20%. This solvates Al, leaving behind a Ni-rich metal. Ni can now be as high as 90%.

To add some further detail to the steps, the addition of sodium hydroxide leaches out Al as shown in Figure 5.3. This is shown here as a stoichiometric reaction but excess sodium hydroxide may be present.

$$2Al + 2NaOH + 6H_2O \rightarrow 3H_{2(g)} + 2Na[Al(OH)_4] \qquad\qquad 70–100\,°C$$

Figure 5.3: Aluminum extraction in Raney nickel production.

Note that in this reaction, which removes aluminum from the alloy, significant amounts of hydrogen are generated, as well as heat. The final catalyst is pyrophoric and must be handled with care and stored in some inert atmosphere, such as water or mineral oil.

5.3 Catalyst fate

Although both nickel and aluminum are not particularly expensive or rare metals, their distribution in the earth is not uniform, and thus, they are tracked by different US government departments, as well as other national ministries [2–5]. As well, as pollution controls in different nations are made stricter, it is often economically more favorable to recover any spent catalyst, rather than dispose of it. Additionally, imports are a continued concern, as import dependency means a feedstock could become unavailable in times of prolonged disruption. This is not solely for the use of either metal as a catalyst. Both find a multitude of other large-scale industrial applications.

When possible, Raney nickel is recovered, since it is possible to reform fresh catalyst, if an existing batch has been deactivated through use. Also, as mentioned, in various states and regions, pollution laws restrict the disposal of a wide variety of metals into landfills or other long-term storage facilities. Nickel has been found, in some situations, to be linked to certain cancers but minimum exposures and oxidation states for such are not fully defined.

References

[1]   Raney, M. Method of producing finely-divided nickel, 1927. U.S. Patent 1,628,190.
[2]   United States Geological Survey. Mineral Commodity Summaries, 2019. Downloadable as: mcs2019.pdf.
[3]   United States Department of Defense, Strategic and Critical Materials 2013 Report on Stockpile Requirements. Downloadable as: Startegic_and_Critical_Materials_2013_ Report_on_Stockpile_Requirements.pdf.
[4]   United States Department of Energy. Critical Materials Strategy, 2011. Downloadable as: DOE_CMS2011_FINAL_Full.pdf.
[5]   British Geological Survey. Natural Environment Research Council. World Mineral Production, 2009–2013. Downloadable as: WMP20092013, British Geological Survey.pdf.

Chapter 6
Ammonia

6.1 Introduction

The production and use of artificial ammonia is one of only a few advances in the history of mankind that has truly and profoundly changed the world. Because of the use of ammonia as a fertilizer, the world population has risen to a total never seen before, over 7.5 billion people. Estimates are that only slightly more than half of the world's population could be fed using fertilizer that was sourced from materials that did not involve artificial ammonia – that synthetic ammonia has allowed the world's population to almost double. The story of the rise of ammonia as a chemical commodity, and its use both as a fertilizer and as a feedstock for military explosives, has been well detailed [1, 2].

The use of ammonia either as fertilizer directly, or as a feedstock for other forms of nitrogen-containing fertilizer, such as ammonium nitrate, is large enough that it is monitored and promoted by different trade organizations [3, 4]. The excess use of it can be a problem, however. This too has been monitored in the recent past [5], and efforts are being made to combat this.

6.2 Reaction chemistry

The reaction chemistry for ammonia production is deceptively simple, an addition re-action between elemental nitrogen and hydrogen, as well as a redox reaction, in which nitrogen is formally reduced and hydrogen oxidized. Figure 6.1 illustrates this.

$$3 \, H_{2(g)} + N_{2(g)} \rightarrow 2 \, NH_{3(l)}$$

Figure 6.1: Ammonia production.

The equilibrium expressed does nothing to explain the reaction conditions, how-ever. High pressure is required, generally 150–200 atm, as well as elevated tempera-ture of approximately 400–500 °C. Importantly, either a promoted iron catalyst or a magnetite catalyst is required, although historically, other metals have been used. The entire operation is referred to as the Haber–Bosch process.

Even with the high pressure and temperature, as well as the use of a catalyst, this reaction does not proceed to completion in a single pass. Unreacted gases are captured and reintroduced to the system to help maximize yields.

https://doi.org/10.1515/9783110542868-006

6.3 Catalyst production

It is noteworthy that the iron catalyst can be promoted with K_2O, CaO, SiO_2, and Al_2O_3. To get to the best catalyst configuration, powderized iron can be burnt and oxidized to give magnetite.

The iron catalysts used in ammonia production require a high surface area, and thus must be produced themselves with care, in order to maximize this. Iron powder is the starting material for the catalyst and is usually produced through the reduction of some magnetite (Fe_3O_4) or other iron-based feedstock. After the iron is of proper particle size, it is oxidized under controlled conditions to obtain a pure magnetite. This powder is then partially reduced, removing some of the oxygen. The resulting particles generally have three layers: an outer iron layer with high porosity, a layer below this that is essentially iron (II) oxide (sometimes still called wüstite), and a core that is magnetite.

In some production methods (usually proprietary), calcium oxides and/or aluminum oxides are introduced and serve as catalyst supports, ensuring that the iron outer layer of the catalyst particles continue to possess the required surface area. To a lesser extent, potassium oxide and silicon oxide can serve this purpose, as all do not react with either nitrogen or hydrogen.

The specificity with which the catalyst for ammonia production is made may at first appear excessive but it should be noted that even with a century of research and experimentation, the production of ammonia does not go to completion in a single pass of reactants in contact with each other in the reaction chamber, as mentioned. Both gases must be recycled into the system multiple times to increase the overall yield of the product.

Also as mentioned, several metals can serve as a catalyst in ammonia production. Ruthenium is used in some operations because it provides a catalyst that can be significantly more active than iron-based ones, which in turn means that the operation can run at somewhat lower temperatures. Called the KAAP process (from KBR Advanced Ammonia Process), Haliburton states at the company website: "KAAP™ synthesis catalyst uses ruthenium as the active ingredient. KAAP™ catalyst is 10–20 times more active than traditional magnetite catalyst. KAAP™ catalyst allows lower synthesis loop pressure than is practical with magnetite catalyst" [6].

The disadvantage of ruthenium as a catalyst is that it is a less common metal than iron, and thus is more costly. Preparation of it involves the reaction of tri-ruthenium dodecacarbonyl ($Ru_3(CO)_{12}$) on a graphite surface. The reaction chemistry is not stoichiometric, and therefore not usually written as a reaction.

6.4 Catalyst fate

Iron-based catalysts tend to last for years, often up to a decade. Past practice has been to dispose of spent catalyst in landfills, simply because the materials are not toxic and iron sources to produce a new catalyst remain plentiful [6]. The newer ruthenium-based catalysts may be more profitable to recycle when they arrive at the end of their usable lifetime, because ruthenium is a much less common element.

References

[1]   Hager, Thomas. The Alchemy of Air, Random House, 2008, ISBN: 978-0-307-35178-4.
[2]   Jeffreys, Diarmuid. Hell's Cartel: IG Farben and the Making of Hitler's War Machine, 2009, ISBN: 978-0-805-09143-4.
[3]   The Fertilizer Institute. Website. (Accessed 26 July 2017, as: https://www.tfi.org/).
[4]   International Fertilizer Association. Website. (Accessed 26 July 2017, as: http://www.fertilizer.org/).
[5]   Hodge, C.A. Pollution Control in Fertilizer Production. ISBN: 978-0471103578.
[6]   Halliburton. Website. (Accessed 26 July 2017, as: http://halliburtoncontracts.com.yeslab.org/kbr/hydroChem/fertSynGas/kaapPlusAmmoniaProcess.html).

Chapter 7
Ammonium sulfate

7.1 Introduction

Ammonium sulfate is a major commodity chemical in the world today. It is used primarily as a fertilizer, as virtually all salts of ammonia are produced on an industrial scale [1–3]. It is favored for alkaline soils. Its production begins with ammonia, which was discussed in Chapter 6. It must also utilize sulfuric acid for the addition of sulfate. Sulfuric acid is discussed later in this volume. Overall, millions of tons are produced annually [4].

7.2 Reaction chemistry

The formation of ammonium sulfate appears to be a simple addition reaction, as shown in Figure 7.1.

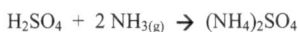

$$H_2SO_4 + 2\,NH_{3(g)} \rightarrow (NH_4)_2SO_4$$

Figure 7.1: Ammonium sulfate production.

The reaction does not however show that water must be added as the reaction proceeds, or that it normally runs at approximately 60 °C, or that a slight excess of sulfuric acid, roughly 2–4%, must be present.

7.3 Catalyst production

The reason the production of ammonium sulfate is included in this volume is that the reaction chemistry, shown in Figure 7.1, must be promoted by the addition of the just-mentioned small amount of sulfuric acid. The amount of the acid is generally between 2% and 4%; it is low enough that one can consider its role to be catalytic. Admittedly, some prefer to call this a promoter and not a catalyst.

Sulfuric acid production is discussed in Chapter 40. As might be imagined, the amount of sulfuric acid used *as an excess* in this reaction is small compared to its use in many other applications.

https://doi.org/10.1515/9783110542868-007

7.4 Catalyst fate

Unlike most catalysts, which are oxidized, reduced, or in some other way deactivated over the course of time, in the process of ammonium sulfate production, simply more sulfuric acid has to be added to the system. Additionally, the cost of sulfuric acid is low enough that it does not need to be recovered for economic reasons. Any recovery is usually driven by local environmental regulations, which are usually in place to prevent the generation of pollution and local environmental degradation.

References

[1]   The Fertilizer Institute. Website. (Accessed 17 February 2021, as: www.tfi.org).
[2]   Fertilizer Industry Roundtable. Website. (Accessed 17 February 2021, as: firt.org).
[3]   International Fertilizer Association. Website. (Accessed 17 February 2021, as: fertilizer.org).
[4]   Sulfuric Acid Today. Website. (Accessed 17 February 2021, as: h2so4today.com).

Chapter 8
Benzene, toluene, xylene (BTX)

8.1 Introduction

The fraction of crude oil that is refined into benzene, toluene, and xylene (BTX) can be isolated by what is called catalytic reforming, using naphtha as a feedstock. The composition of naphtha changes to some degree with every batch that is introduced for the process. All three of these target molecules are clear liquids that have found use as organic solvents. But all are also transformed into several other commodity chemicals that have a higher value, often with the aim of producing monomers for plastics.

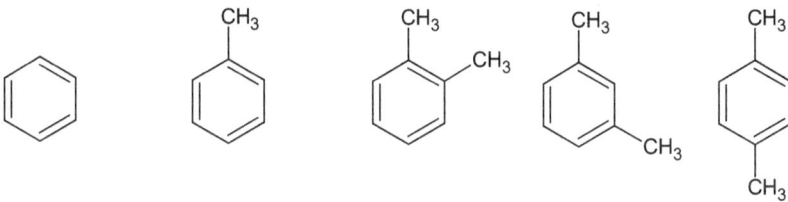

Figure 8.1: The BTX molecules.

Figure 8.1 shows the Lewis structures of the three molecules and illustrates xylene as it exists in its three isomers. Undoubtedly, the *para* isomer is by far the most useful, as it becomes a feedstock for what is one of the high volume polymers, polyethylene terephthalate (PETE). Both molecules are shown in Figure 8.2.

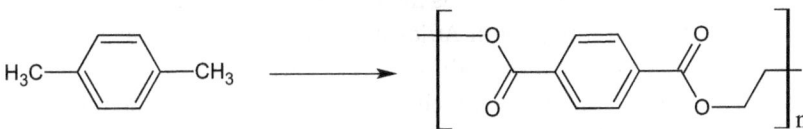

Figure 8.2: *para*-Xylene and the PETE repeat unit.

While the other isomers of xylene are not without value, they have no large-scale use to compare with that of *para*-xylene. Table 8.1 lists the commodity chemicals that are routinely made from the three components of BTX. Other products can be made as well [1].

https://doi.org/10.1515/9783110542868-008

Table 8.1: Commodities produced from BTX.

BTX component	Commodity chemical	Polymer
Benzene	Cumene	Polycarbonate, phenolic resin
	Cyclohexane	Nylon-6
	Styrene	Polystyrene (PS)
Toluene	Motor fuel additive	
	Toluene diisocyanate (TDI)	Polyurethanes
ortho-Xylene	Phthalic anhydride	
para-Xylene	Terephthalic acid	Polyterephthalate (PETE)

8.2 Reaction chemistry

It is difficult to represent a reaction form for the production of BTX from naphtha because the starting material is such a complex mixture of compounds. However, the process must first run through what is called a catalytic reformer before the BTX is separated and purified.

The catalyst required in the catalytic reformer is very often platinum or rhenium on a silica support or bed, or silica–alumina support. More recently, some zeolites have been utilized but are often proprietary to the company that uses them [2, 3].

8.3 Catalyst production

Platinum and rhenium are both rare metals mined in a limited number of locations throughout the world, including South Africa and Russia. The Bushveld Igneous Complex under South Africa is currently the largest producer.

A secondary source of rare metals, including these two, is the electrolytic refining of copper or lead, and the recovery of what is called anode mud, anode slime, or anode sludge. The production of precious metals like platinum and rhenium, as well as other rare elements, is tracked by national governments because of their scarcity and defined uses [4–6]. Platinum is often sold for use as a catalyst in powder form, mixed with carbon black, as shown in Figure 8.3.

A small amount of chlorine is added to such catalysts to promote activity. In this area, the use of less chlorine remains an area of importance, in part because of economic factors – finding the least expensive catalyst or catalyst mix – for any operation in which a platinum catalyst is required [7].

Figure 8.3: Platinum black.

8.3.1 Aluminum oxide, aka alumina, aka aluminum (III) oxide

As a crystal, with only minor impurities, this material is a gemstone, such as sapphire, and sometimes as ruby. As a fine powder, it is a white, abrasive material, which can function as a catalyst because of its high surface area.

At approximately 1,100 °C, the following reaction $2Al(OH)_3 \rightarrow Al_2O_3 + 3H_2O_{(g)}$ produces high-grade alumina. Once again, this compound is produced in large quantities, with catalysts being a relatively minor use, especially when compared to its use in producing elemental aluminum metal [8–10].

8.4 Catalyst fate

The precious metal or metals used in BTX formation is expensive enough that even when it is deactivated it is collected so that it may be regenerated or reduced to the metal. Platinum and rhenium are rare enough and valuable enough that they are tracked annually in the USGS Mineral Commodity Summaries, as well as other agencies internationally [4–6]. These metals can be deactivated by the slow accumulation of coke impurities during the process cycle, or because of the slow, long-term loss of the chloride as the process runs.

In general, extending the lifetime of any catalyst is economically favorable but when a catalyst is completely expended, its fate is determined by its composition. As mentioned, platinum and rhenium are recovered in some way, often by chemical or electrochemical reduction. Alumina, silica, clays, and zeolites that are made inexpensively enough can be discarded or used in other applications where some inert filler is useful. If the components – usually the metal within such a clay – are expensive enough, the catalyst can be recovered for reclamation of its components.

References

[1]   American Chemistry Council. Website. (Accessed 6 October 2020, as: plastics. americanchemistry.com/default.aspx).
[2]   Clariant. Website. (Accessed 3 September 2020, as: https://www.clariant.com/en/Business-Units/Functional-Minerals/BTX-Catalysts).
[3]   European patent. EP1232231A1. Process for BTX Purification.
[4]   U.S.G.S. Mineral Commodity Summaries 2020. Downloadable document at: usgs.gov/centers/nmic/mineral-commodity-summaries.
[5]   Minerals and Metals Facts. Website. (Accessed 6 October 2020, as: nrcan.gc.ca/our-natural-resources/minerals-mining/minerals-metals-facts/20507).
[6]   Collection Mineral Surveys. Website. (Accessed 6 October 2020, as: www.gov.uk/government/collections/minerals).
[7]   U.S. Patent. US5516741A. Reduced chlorine containing platinum catalysts.
[8]   The Aluminum Association. Website. (Accessed 26 September 2020, as: aluminum.org).
[9]   European Aluminum. Website. (Accessed 26 September 2020, as: European-aluminum.eu).
[10] Bhasin, M.M., McCain, J.H., Vora, B.V., Imai, T., and Pujado, P.R. Dehydrogenation and oxydehydrogenation of paraffins to olefins. Applied Catalysis A: General, 221, 1–2, (2001), 397–419.

Chapter 9
Butadiene

9.1 Introduction

The refining of crude oil produces what is called a C4 fraction, which is predominantly butanes. The two major products in this fraction which are used industrially are n-butane – for the production of butadiene – and isobutene. Some 1,3-butadiene is produced in this manner but most of them are made from the catalytic dehydration of n-butane. The main use of this monomer continues to be for the production of synthetic rubber.

The rise of the use of 1,3-butadiene – often simply called butadiene – has its origins in, and is directly coupled to, the Allied war effort in the Second World War. The Empire of Japan had taken much of the area in the world where natural rubber was grown and harvested as an exudate of the *Hevea brasiliensis* trees and was able to deny access to it to the nations fighting against it, the Allies. Thus, a search was made for some artificial replacement, and 1,3-butadiene was found to be an acceptable monomer for rubber in several applications. The source for it, crude oil, was something the United States had in abundance during the war. After the conclusion of the war, the technical know-how had been developed to keep butadiene production high, since the number of applications for which synthetic rubber could be used continued to grow.

9.2 Reaction chemistry

Figure 9.1 shows the basic chemistry for the dehydrogenation of n-butane. The coproduct is elemental hydrogen.

Figure 9.1: Production of 1,3-butadiene.

Although the dehydrogenation of butane seems to be a logical starting point to produce butadiene, the coupling of ethanol is also a means by which this unsaturated hydrocarbon can be produced and has become a significant use for ethanol [1–4]. The reaction chemistry is shown in Figure 9.2. Notice that both water and elemental hydrogen are the coproducts.

The dehydrogenation of n-butane and the ethanol coupling are large-scale procedures because the butadiene product is routinely used in the just-mentioned production of synthetic rubber. Figure 9.3 compares the two monomers: butadiene for

https://doi.org/10.1515/9783110542868-009

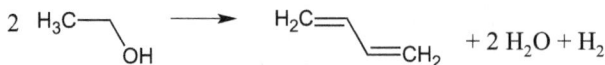

Figure 9.2: Ethanol coupling.

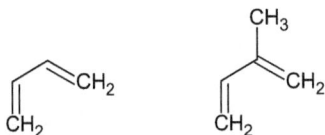

Figure 9.3: Lewis structures of butadiene and isoprene.

synthetic rubber and isoprene for natural rubber. The rather obvious difference is the single methyl group that is part of isoprene.

9.3 Catalyst production

The Houdry Catadiene process – also called the Oxo-D™ process – is an established method in which butane is dehydrogenated over chromia and alumina, and was first put into large-scale practice in the late 1930s. Temperatures must still be elevated.

9.3.1 Chromium (III) oxide, aka chromia

Because the compound is green, chromia also finds a significant use as a paint pigment [5–8]. Indeed, the first preparation of chromia in the nineteenth century was a proprietary process precisely because of the economic value of it as a green pigment. The reaction chemistry $Na_2Cr_2O_7 + S \rightarrow Cr_2O_3 + Na_2SO_4$ to form it at elevated temperature is an established means to produce this material. Other methods involve the decomposition of other chromium salts.

The source for chromia is the mineral chromite, which can be a mixed iron, magnesium, and chromium oxide – represented as $Mg_xFe_yCr_2O_4$. Purification is required to obtain usable chromia, which is why the above method is often preferred in arriving at Cr_2O_3.

9.3.2 Aluminum oxide, aka alumina, or aluminum (III) oxide

The production of alumina is discussed in Chapter 8, since it is also used in benzene, toluene, and xylene production.

Ethanol dehydration and coupling utilizes metal oxides as a catalyst as well but occurs using a wider variety than just alumina or chromia.

9.4 Catalyst fate

Both alumina and chromia are produced on large enough scales that their recovery is not always economically advantageous. Thus, they can be disposed of when the catalyst has been used long enough that its activity has decreased.

References

[1] American Coalition for Ethanol. Website. (Accessed 26 September 2020, as: ethanol.org).
[2] Ethanol, Natural Resources Canada. Website. (Accessed 26 September 2020, as: rncan.gc.ca).
[3] iEthanol. Website. (Accessed 26 September 2020, as: iethanol.eu).
[4] Ethanol Europe. Website. (Accessed 26 September 2020, as: eerl.com).
[5] Canadian Paint and Coatings Association. Website. (Accessed 26 September 2020, as: canpaint.com).
[6] Color Pigments Manufacturers Association. Website. (Accessed 26 September 2020, as: www.pigments.org).
[7] CEPE. Website. (Accessed 26 September 2020, as: cepe.org).
[8] International Paint and Printing Ink Council. Website. (Accessed 26 September 2020, as: worldcoatingscouncil.org

Chapter 10
Caprolactam

10.1 Introduction

Caprolactam is now a major commodity chemical, with production runs annually of over 4 billion kilograms (roughly 5 million tons). It is almost entirely used as a precursor for the formation of nylon-6, although a smaller amount is used in other plastics, and a very small amount is used as a chemical intermediate. In Table 10.1, the following companies are noted as major producers of the material:

Table 10.1: Caprolactam producers.

Name	Amount produced	Other products	Website
BASF		Amines, plastics	basf.com
China Petrochemical			
China Petroleum & Chemical Corp. (Sinopec Ltd.)			www.sinopecgroup.com
Honeywell International, Inc.			www.honeywell.com
Kuibyshevazot Ojsc	Sold as 25 kg bags, flake	Fertilizers, fibers	https://www.kuazot.ru/en/products/kaprolaktam/
Lanxess AG		Plastics, glass fiber	https://techcenter.lanxess.com/scp/emea/en/products/intem/article.jsp?docId=61971
Royal DSM N.V.		Sold caprolactam unit to Highsun Holdings, 2018	
Shandong Haili Chemical Industry Co., Ltd.	200 K ton/year		www.jxjyjl.com
Sumitomo Chemical Company, Ltd.		Petrochemical and plastics	https://www.sumitomo-chem.co.jp/english/products/detail/en_a02005.html
UBE Industries	Sold as 25 kg bags, flake	Rubber, plastics, machinery	https://www.ube.com/contents/en/chemical/caprolactam/caprolactam.html

https://doi.org/10.1515/9783110542868-010

Note that there are other producers as well, for some of whom caprolactam is not their major product but rather a smaller one. The market for this product is a mature one.

10.2 Reaction chemistry

Caprolactam is routinely produced from cyclohexanone in a two-step process in which the nitrogen atom is inserted, as shown in Figure 10.1. The final step is an example of what is called a Beckmann rearrangement, which can be catalyzed by a variety of acids [1]. Sulfuric acid is often used because it is not expensive and it produces ammonium sulfate as a by-product, when ammonia is used to neutralize it. This ammonium sulfate finds use as a fertilizer, and thus becomes a second revenue stream, as opposed to a waste stream, when caprolactam is produced.

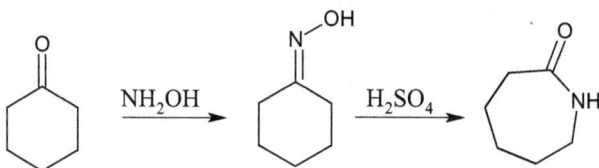

Figure 10.1: Caprolactam production.

10.3 Catalyst production

Sulfuric acid is treated later in this book and remains the largest commodity chemical produced in the world. Its use in caprolactam production is minor with the bulk of it being used in the production of phosphate fertilizers.

10.4 Catalyst fate

Sulfuric acid is inexpensive enough that it can be disposed of but companies often recover it, as well as other acids that might be used because of the penalties associated with its unregulated disposal.

Reference

[1]   CN 109718828. "Microsphere silicate-1 molecular sieve catalyst, preparation method thereof and method for preparing caprolactam" Cheng, Shibiao; Xie, Li; Zhang, Shuzhong; Mu, Xuhong; Zong, Baoning.

Chapter 11
Chlorine

11.1 Introduction

The production of elemental chlorine is another absolutely enormous industrial process, conducted in numerous places in many countries throughout the world. But it is one that has had both positive and negative uses in the past century. Indeed, the large-scale production of chlorine has had profound influences on life in the twentieth and twenty-first centuries. It is remembered as the first poison gas used during the First World War. On a much more positive note, it is widely used today as an inexpensive means of cleaning water of harmful bacteria.

The production of chlorine continues to be part of a three-product process, the other two being sodium hydroxide and elemental hydrogen. Its scale is large enough, and it is economically advantageous enough that several national or international trade organizations are devoted to the production and positive use of this element [1–6].

11.2 Reaction chemistry

The basic chemistry through which chlorine is produced is illustrated in Figures 11.1 and 11.2. Figure 11.1 is referred to as the chlor-alkali process and is electrolytic in its execution. In this process, chlorine is oxidized from brine solution, hydrogen is reduced, and sodium hydroxide is also formed, which lasts through neither oxidation nor reduction. Throughout history sodium hydroxide has been the most valuable product economically.

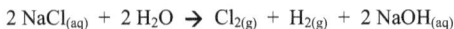

$$2\,NaCl_{(aq)} + 2\,H_2O \rightarrow Cl_{2(g)} + H_{2(g)} + 2\,NaOH_{(aq)}$$

Figure 11.1: Chlor-alkali process.

The process shown in Figure 11.2, a Downs' cell, is a pyrometallurgical one in which molten salt is spilt into its component elements. One large faction of the cost involved in this is the generation of the power needed to keep the temperature of the reaction sufficiently high.

$$2\,NaCl_{(l)} \rightarrow 2\,Na_{(l)} + Cl_{2(g)}$$

Figure 11.2: Downs' cell chlorine production.

https://doi.org/10.1515/9783110542868-011

Note that the sodium is a molten metal when it forms as a product.

Both of these processes are effective at producing elemental chlorine. But both are not processes that require a catalyst.

An older, now displaced process for the production of chlorine is known as the Deacon process, and routinely utilized chlorine that was recovered during the large-scale production of chlorinated organic compounds. The reaction chemistry, in Figure 11.3, is straightforward, involving the reduction of oxygen to induce the oxidation of chlorine.

$$O_{2(g)} + 4\,HCl \rightarrow 2\,H_2O + 2\,Cl_2$$

Figure 11.3: The Deacon process reaction.

Although this process has been replaced by the chlor-alkali process, to give credit where it is due, the Deacon process worked and utilized a chlorine feedstock that otherwise would have been wasted. The process does require a catalyst [7], which were originally copper based, such as $CuCl_2$. As the process was exploited to a greater extent, from Deacon's original work in the 1870s, it was found that chromium-based catalysts as well as ruthenium-based salts were also effective. As well, the reaction required elevated temperatures – 400–450 °C – for good product formation.

11.3 Catalyst production

Copper chloride – more properly, copper (II) chloride or cupric chloride – one of the catalysts that has proven useful in the older Deacon process, is produced rather ironically by the addition of elemental chlorine to copper. Figure 11.4 shows the basic chemistry.

$$Cl_{2(g)} + Cu_{(s)} + 2\,H_2O \rightarrow CuCl_2(H_2O)_2$$

Figure 11.4: Copper chloride production.

The hydrated product is then dehydrated to produce a pure copper chloride salt. While this can be produced by other means, this has been the commercially viable way to produce copper chloride for decades.

11.4 Catalyst fate

Since the Deacon process is now considered outdated, compared to the electrolytic processes by which chlorine is produced, catalyst recovery is no longer a major concern.

References

[1]   American Chemistry Council. Website. (Accessed 19 December 2019, as: https://chlorine.
       americanchemistry.com/Chlorine/).
[2]   The Chlorine Institute. Website. (Accessed 19 December 2019, as: https://www.chlorineinsti
       tute.org/).
[3]   Pamphlet 10) North American Chlor-Alkali Industry Plants and Production Data Report- 2017 –
       November 2018.
[4]   Website. (Accessed 19 December 2019, as: https://bookstore.chlorineinstitute.org/chlorine-
       manufacturing-and-industry-data.html).
[5]   Eurochlor. Website. (Accessed 19 December 2019, as: https://www.eurochlor.org/).
[6]   World Chlorine Council. Website. (Accessed 19 December 2019, as: https://worldchlorine.org/).
[7]   Borkowski, W.A. Preparation of chlorine from hydrogen chloride. U.S. Patent 3,542,520.
       (Accessed 16 April 2020).

Chapter 12
Cumene

12.1 Introduction

Cumene is another of the bulk, commodity chemicals ultimately sourced from petroleum, and ultimately transformed into some other chemical, rather than being
sold as is. Because it is so routinely transformed into acetone and phenol, there is
seldom mention of it in the popular press, and even little in chemistry textbooks that
focus on organic chemistry [1]. Its synthesis is from two other commodity chemicals
of lesser value: benzene and propylene.

12.2 Reaction chemistry

The basic reaction chemistry for the production of cumene starts with benzene and propylene, and some Lewis acid as a catalyst. The reaction is classified as a Friedel–Crafts
alkylation, since benzene is substituted in the process. The choice of Lewis acid as
catalyst can be wide, since many are effective. Phosphoric acid, a large volume commodity chemical, can be used. But likewise, aluminum halides serve this purpose as
well. Figure 12.1 shows the basic reaction chemistry.

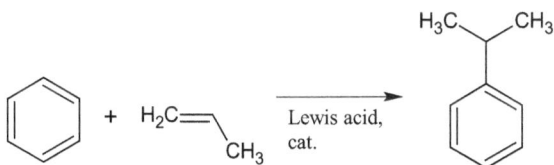

Figure 12.1: Formation of cumene.

Besides the catalyst, reaction conditions are an elevated temperature (ca. 250 °C)
and pressure (ca. 30 atm), making the formation of a gas-phase reaction. Most cumene is not isolated, however. It is converted directly to phenol and acetone – this
latter conversion being titled the cumene process. Overall, this represents a conversion of two organic starting materials to two organic products of higher economic
value. Figure 12.2 shows the conversion to phenol and acetone.
    The oxygen in Figure 12.2 simply comes from the air.

https://doi.org/10.1515/9783110542868-012

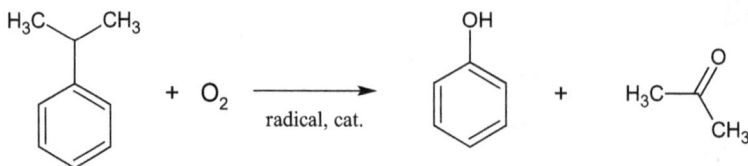

Figure 12.2: The cumene process.

12.3 Catalyst production

Since the production of cumene is widespread, and several different companies produce it in some proprietary fashion, we will treat two of the more common catalysts that have been used on a large scale: aluminum chloride and phosphoric acid.

12.3.1 Aluminum chloride

The compound $AlCl_3$ is produced on a large scale by the direct combination of elemental chlorine and powdered aluminum but it can also be produced starting with hydrogen chloride. The reaction must be run at elevated temperatures, generally 650–700 °C, and is highly exothermic. Figure 12.3 illustrates both addition reactions.

$$2\ Al_{(s)} + 3\ Cl_{2(g)} \rightarrow 2\ AlCl_{3(s)}$$

or

$$2Al_{(s)} + 6HCl \rightarrow 2\ AlCl_{3(s)} + 3\ H_{2(g)}$$

Figure 12.3: Production of aluminum chloride.

Additionally, copper (II) chloride can be used as a starting material and combined with elemental aluminum, resulting in aluminum chloride and copper metal, in a single displacement reaction. This does not appear to be as widely used a process as those two shown in Figure 14.2, however.

12.3.2 Phosphoric acid

The production of phosphoric acid is an enormous one, as this is one of the top ten commodity chemicals produced worldwide. Although several methods produce the acid, what is called the wet process is by far the most common. Generally, the production is by the treatment of phosphate rock by sulfuric acid. The simplified reaction chemistry is shown in Figure 12.4.

$$H_2SO_4 + Ca_3(PO_4)_2 \rightarrow CaSO_4 + H_3PO_4$$

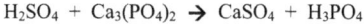

Figure 12.4: Phosphoric acid production.

The coproduct, usually called gypsum, is separated from the main product by filtration, since it is not soluble in aqueous medium, and H_3PO_4 is isolated. Since most phosphoric acid is used as fertilizer, the purity is not often presented in terms of acid but rather in terms of P_2O_5 produced. Routinely, 42–45% P_2O_5 is isolated. Since phosphate rock is a common starting material, a step is also involved that removes impurities in the rock, routinely by means of filtration of insoluble materials [2].

12.4 Catalyst fate

The two catalysts discussed here are common enough, and made from inexpensive enough materials, that efforts to recover them are based on the economics of their disposal (how much disposal costs versus how much recovery will save on production costs).

Importantly, it was mentioned that a wide variety of Lewis acids can be used in the role of catalyst here. Among the more recent catalysts are certain zeolites. These are sometimes expensive enough catalysts that their recovery is warranted.

References

[1]   BuyersGuide.com. Website. (Accessed 19 December 2019, as: https://www.buyersguide chem.com/chemical_supplier/cumene).
[2]   United states Geological Survey. Mineral Commodity Summaries, 2019. Downloadable as: mcs2019_all.pdf.

Chapter 13
Cyclohexane

13.1 Introduction

Cyclohexane finds large-scale use as a nonpolar solvent, and as a solvent it facilitates a wide number of different reactions. Interestingly, while it is a solvent, it is also a precursor chemical for the production of adipic acid and caprolactam. Thus, cyclohexane ultimately is a precursor for the manufacture of nylon. Indeed, approximately 90% of cyclohexane is produced to make these nylon precursor materials.

Also, cyclohexane can be selectively oxidized to produce what is called "KA oil," an abbreviation for "ketone–alcohol," as shown in Figure 13.1. This is because the two products from this selective oxidation are cyclohexanone and cyclohexanol.

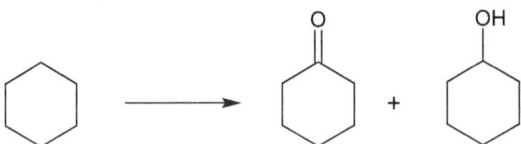

Figure 13.1: Production of KA oil.

Several major producers of cyclohexane are shown in Table 13.1, although the list is not exhaustive. It is not surprising that these companies are major producers of a wide variety of other organic materials as well.

Table 13.1: Cyclohexane producers.

Name	Annual capacity (tons)	Comments, other products
BASF	50,000	Produces large variety of organic chemicals
CEPSA	180,000	Oil and gas, headquartered in Spain
Chevron Phillips Chemical	Not disclosed	Claims to be world's largest producer
ExxonMobil	Not disclosed	Petroleum–gasoline as well
Huntsman	362,000	Amines and surfactants as well

From Refs. [1–5].

As can be imagined with several companies competing for market share, the catalysts used for the production of cyclohexane, or at least the details of their manufacture and surface area, can be proprietary. But quoting one website:

https://doi.org/10.1515/9783110542868-013

From Chevron Phillips:

> All cyclohexane is produced in benzene hydrogenation units. In the process, high-purity benzene feed and purified hydrogen (typically recovered from reformers and ethylene crackers) are brought to reaction temperatures and charged to the reactor. The conversion of benzene to cyclohexane is stoichiometric and almost complete; finished cyclohexane typically contains less than 50 ppm of benzene. A small amount of lower purity cyclohexane is recovered from petroleum streams by fractionation and extraction. [3]

Although a catalyst is not mentioned, in general, Raney nickel or platinum can be used for the hydrogenation of benzene.

13.2 Reaction chemistry

The production of virtually all cyclohexane is from benzene, which is itself separated and distilled from crude oil. This reaction is large enough that more than 10% of the benzene produced annually is used for it. Figure 13.2 shows the simplified reaction chemistry.

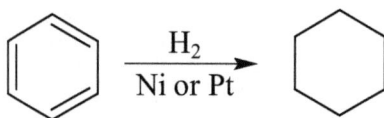

Figure 13.2: Hydrogenation of benzene to produce cyclohexane.

As mentioned, more than one metal in the solid phase can act as the hydrogenation catalyst for this reaction, and thus the reaction is one that uses a heterogeneous catalyst. Corporations that use nickel-based materials often do so because it is inexpensive. Platinum is used for the simple reason that it has been found to work extremely well. Raney nickel is also used extensively and has been discussed in some detail in Chapter 5.

13.3 Catalyst production

The mining and production of metals such as nickel and platinum are tracked by different nations, since these commodities are used in several other applications as well, including their defense departments or ministries. In the United States, the *USGS Mineral Commodity Summaries*, an annual publication, tracks the production of nickel, as well as of the platinum group metals (the PGM) [6]. The PGM indicates that the United States is over 60% dependent on imports of both. It states about the uses of nickel:

Approximately 48% of the primary nickel consumed went into stainless and alloy steel products, 40% into nonferrous alloys and superalloys, 8% into electroplating, and 4% into other uses. [6]

Thus, the bulk of nickel is not produced for use as catalysts; rather, it finds use in the production of high-performance steels. Additionally, in the past five decades, nickel has also found increasing use in small denomination coins of many nations.

And of platinum, the *Mineral Commodity Summaries* state:

The leading use for PGMs was in catalytic converters to decrease harmful emissions from automobiles. PGMs are also used in catalysts for bulk-chemical production and petroleum refining; in electronic applications, such as in computer hard disks, in multilayer ceramic capacitors, and in hybridized integrated circuits; in glass manufacturing; in jewelry; and in laboratory equipment. Platinum is used in the medical sector; platinum and palladium, along with gold-silver-copper-zinc alloys, are used as dental restorative materials. [6]

As well, the production of copper and its refining to high purity entails the production of a somewhat smaller amount of PGM, as a by-product. It can be seen that the use of platinum as a catalyst is one of its major functions, although it can be used in a variety of reactions – not just cyclohexane production – all of which require some catalyst in order to reach completion.

Beyond the mining and purification of these elemental metals, they must be further shaped to maximize surface area. The process by which this is done for platinum results in what is called platinum black, simply because the material has a smooth, black look, despite it having a varied surface at the microscopic level. It can be represented in reaction form as shown in Figure 13.3, although it is not shown as a stoichiometric reaction. The temperature is routinely at 500 °C, since the sodium nitrate is in the melt. As well, chloroplatinic acid can serve as the starting material in this separation and isolation [7].

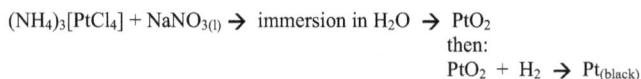

$$(NH_4)_3[PtCl_4] + NaNO_{3(l)} \rightarrow \text{immersion in } H_2O \rightarrow PtO_2$$
$$\text{then:}$$
$$PtO_2 + H_2 \rightarrow Pt_{(black)}$$

Figure 13.3: Production of platinum black.

13.4 Catalyst fate

Catalyst recovery, especially of platinum, is driven by the economics of the catalyst's cost. Since metal catalysts for this reaction are all classed as heterogeneous catalysts, their recovery from solution is generally a matter of distillation, to separate them from any liquid or liquid-phase products. In all cases, platinum is recovered when at all possible.

References

[1]   BASF, website. (Accessed 6 August 2019, as: www.basf.com).
[2]   CEPSA, website. (Accessed 16 August 2019, as: https://www.cepsa.com/en/the-company).
[3]   Chevron Phillips Chemical, website. (Accessed 16 August 2019, as: http://www.cpchem.com/en-us/Pages/default.aspx).
[4]   ExxonMobil, website. (Accessed 16 August 2019, as: https://corporate.exxonmobil.com/).
[5]   Huntsman, website. (Accessed 16 August 2019, as: https://www.huntsman.com/corporate/a/Home).
[6]   United States Geological Survey, Mineral Commodity Summaries 2018, downloadable as: https://www.usgs.gov/centers/nmic/mineral-commodity-summaries.
[7]   Silverwood, I.P., and Armstrong, J. Surface diffusion of cyclic hydrocarbons on nickel. Surface Science, 674, (2018), 13–17.

Chapter 14
Ethanol

14.1 Introduction

The production of ethanol has a truly ancient history as a chemical process, always in the form of creating a common and well-known alcoholic beverage – wine or beer throughout most of history – made from the fermentation of grains or fruits. In this type of formation of ethanol, the tiny organism yeast can be considered the catalyst which turns starches into ethanol, simply because a very small amount of yeast produces a much larger amount of this product.

Much more recently, ethanol has been produced by the hydration of ethylene, the latter a product formed from the distillation of crude oil. This process requires a catalyst that is not biotechnological in nature, phosphoric acid supported on some medium.

The manufacture of yeast is a large enough industrial operation that trade associations exist to promote its uses [1, 2]. Likewise, the production of phosphoric acid and of supporting media are large enough sectors of the chemical industry that they too have trade associations associated with them [3, 4]. The production of phosphate rock, from which phosphoric acid is produced, is tracked annually by the United States Geological Survey in its *Mineral Commodity Summaries* [5]. Most of the phosphoric acid produced, though, is used for fertilizer.

14.2 Reaction chemistry

14.2.1 From starches

The reaction chemistry for the production of ethanol from any starch source can be represented easily, as shown in Figure 14.1. One mole of a sugary or starchy material, such as glucose, is broken down biologically into both ethanol and carbon dioxide.

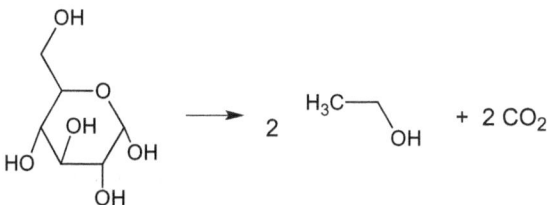

Figure 14.1: Production of ethanol from sugar.

https://doi.org/10.1515/9783110542868-014

This can also be represented as the following stoichiometric equation:

$$C_6H_{12}O_6 \rightarrow 2\ C_2H_5OH + 2CO_2$$

Throughout much of history, the highest percentage of ethanol produced in any fermentation is roughly 15%, the point at which the yeast dies in the ethanol-containing product. Also, throughout much of history, yeast was unknown. Wild, airborne yeasts that blew into fermentation vats caused the fermentation to take place but were not detected by or known to the brewers. Perhaps the earliest example of a recipe or law that covers the production of ethanol, and that does not include yeast, is the Bavarian German Reinheitsgebot of 1516.

> We hereby proclaim and decree, by Authority of our Province, that henceforth in the Duchy of Bavaria, in the country as well as the cities and marketplaces, the following rules apply to the sale of beer: . . . the only ingredients used for the brewing of beer must be Barley, Hops, and Water.

Yeast is not mentioned in this purity law, yet is essential to the production of beer. It was, however, not known at the time the law was written.

One advantage of the production of ethanol from a plant source, as opposed to ethylene, is that the feedstock is renewable. Additionally, the use of yeast is one that has been proven effective by nature, and thus needs little by way of improvement.

14.2.2 From ethylene

Shell Oil first produced ethanol from ethylene shortly after the Second World War. Shell and several other companies use variations of a system that requires phosphoric acid (H_3PO_4) supported on a nonreactive surface such as silica or diatomaceous earth. The simplified reaction chemistry is shown in Figure 14.2, without mention of the catalyst. Water is added as high-pressure steam at approximately 300 °C. The reaction is not a stoichiometrically exact one; and it has been found that the ratio of ethylene feed to steam must be slightly higher than 5:1.

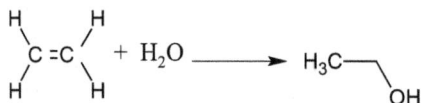

Figure 14.2: Ethanol produced from ethylene.

Although phosphoric acid has been used widely, it is not the only acid that can function as a catalyst in this hydration.

The driving force for which feedstock is to be used in ethanol production is often an economic one, which means the process is less expensive. In the United

States, subsidies for growing corn have affected the price of corn, and thus of the starch used to produce ethanol.

14.3 Catalyst production

14.3.1 Yeasts

Numerous variations on yeast exist, yet relatively few companies produce yeast on a large scale. Those listed in Table 14.1 may sell yeast to bakeries as well as breweries; and indeed, some companies may keep their yeast cultures as trade secrets.

Table 14.1: Yeast producing corporations.

Name	Product(s)	Comments	Website
AB Vista	Markets "Vistacell"	Major producer of animal feed enzymes	https://www.abvista.com
AB Mauri	"Fleischman's" yeast products		http://abmna.com/
Angel Yeast Co., Ltd.		In China and Singapore	http://en.angelyeast.com/
DSM		Wide range of products, including food supplements	http://www.dsm.com/markets/foodandbeverages/en_US/home.html
GB Plange		Acquired by AB Mauri in 2014	
Lallemand		Producing yeasts since 1923	http://www.lallemand.com/
Lesaffre Group		Products in healthcare as well	http://www.lesaffre.com/
Wyeast	Markets multiple brewing/baking yeasts		http://www.wyeastlab.com/yeast-fundamentals

From Refs. [6–15].

14.3.2 Phosphoric acid

There is more than one large scale means by which phosphoric acid can be produced. One major method, called the wet process, involves the reaction of sulfuric acid, as shown in Figure 14.3, with hydroxyapatite. Other phosphate-containing minerals can be used as well.

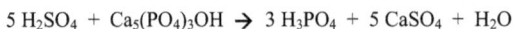

$$5\ H_2SO_4 + Ca_5(PO_4)_3OH \rightarrow 3\ H_3PO_4 + 5\ CaSO_4 + H_2O$$

Figure 14.3: Phosphoric acid production.

This reaction does not routinely run to completion. Rather, it produces a solution that can be concentrated for use.

A second technique, which utilizes a set of distinct reactions, can be used to produce a purer phosphoric acid, including one that is a food grade (food-grade phosphoric acid can be found in some popular soft drinks). This grade of phosphoric acid has E338 as an additive number in Europe. It is difficult to represent simply with equations, since the reactions do not tend to be stoichiometric. But the steps can be broadly listed:

1. Coke and phosphate ore are reacted to produce elemental phosphorus.
2. Silica is then added to the reduced phosphorus to create $CaSiO_4$, a slag.
3. The purer elemental phosphorus is heated in air to produce P_2O_5.
4. Finally, P_2O_5 reacts with water to produce the desired H_3PO_4.

This technique tends to be the more expensive of the two, at least when the use of the product is for the subsequent production of ethanol. But run on large scales, both are effective means of producing phosphoric acid.

14.3.3 Silica

Silica is essentially quartz, and thus can be mined, and then ground to fine enough particles that it can be used as a support for numerous reactions, including the production of ethanol. Production of this grade of silica involves sieving the crushed material until the particles are of the desired size for use.

14.3.4 Diatomaceous earth

Diatomaceous earth is fossilized diatoms and can be 90% silica. Because of its porosity and high surface area, it has a very low density. Its feel is rough and abrasive, and the material is used in a variety of ways when materials require polishing (there is a grade that can be used in small amounts in toothpaste). The high porosity of diatomaceous earth makes it an excellent catalyst support, since it helps maximize the area of the catalyst. Diatomaceous earth is routinely mined, then cleaned of contaminants.

It is then ground to powder to maximize its own surface area. This is then mixed with any catalyst for which it must serve as a support.

14.4 Catalyst fate

All of the catalyst materials for ethanol production are themselves very inexpensive. Yeast can die in ethanol it produces, and thus has to be replaced. However, yeast cultures can be grown readily and can be kept for long periods of time to be used in some future production run. Companies that produce and sell yeast do keep batches of the organism so that they can continually breed or produce more.

The cost of phosphoric acid as well as that of silica and diatomaceous earth is generally low enough that they too can be replaced as needed and are not usually recycled.

References

[1]   CEFIC Sector Groups, PAPA – Phosphoric acid and phosphates association. Website. (Accessed 21 June 2019, as: https://specialty-chemicals.eu/papa/).
[2]   Association of Synthetic Amorphous Silica Producers. Website. (Accessed 24 June 2019, as: https://www.asasp.eu/).
[3]   Confederation of European Yeast Producers. COFALEC. Website. (Accessed 21 June 2019, as: https://www.cofalec.com/).
[4]   European Association for Specialty Yeast Products. Website. (Accessed 24 June 2019, as: http://www.yeastextract.info/about).
[5]   U.S. Geological Survey. Mineral Commodity Summaries, 2019, downloadable as: mcs2019.
[6]   AB Vista. Website. (Accessed 19 February 2018, as: https://www.abvista.com).
[7]   AB Mauri. Website. (Accessed 19 February 2018, as: http://abmna.com/).
[8]   Angel Yeast Co., Ltd. Website. (Accessed 3 May 2018, as: http://en.angelyeast.com/).
[9]   DSM. Website. (Accessed 19 February 2018, as: http://www.dsm.com/markets/foodandbeverages/en_US/home.html).
[10] Lesaffre Group. Website. (Accessed 19 February 2018, as: http://www.lesaffre.com/).
[11] Lallemand. Website. (Accessed 19 February 2018, as: http://www.lallemand.com/).
[12] Explore Yeast. Website. (Accessed 19 February 2018, as: http://www.exploreyeast.com/).
[13] Yeast Production. Website. (Accessed 19 February 2018, as: https://www3.epa.gov/ttnchie1/ap42/ch09/final/c9s13-4.pdf).
[14] Commercial Yeast Production. Website. (Accessed 19 February 2018, as: http://www.lallemand.com/wp-content/uploads/2011/07/Prod-Proc-11.pdf).
[15] Wyeast. Website. (Accessed 19 February 2018, as: http://www.wyeastlab.com).

Chapter 15
Ethylene

15.1 Introduction

The large-scale production and use of ethylene in the twentieth century is that of a chemical that has changed the world profoundly. Annually, over 150 million tons are produced, and the vast majority of ethylene for further use in the production of different types of polyethylene [1]. The reaction of ethylene to polyethylene can be shown in a simplified way, as in Figure 15.1. In this chapter, however, we will not discuss the catalysts needed to affect the polymerization but rather that needed to produce ethylene itself.

$$n \ H_2C{=}CH_2 \longrightarrow$$

Figure 15.1: Polyethylene production from ethylene.

15.2 Reaction chemistry

It is difficult to show a balanced chemical equation for the production of ethylene from petroleum, since it is separated from a mixture of many hydrocarbons.

The dehydration of ethanol to produce ethylene is another established means of making this commodity chemical, and one that is seeing a resurgence in recent years, as the growth in the production of bioethanol has continued. This dehydration is shown in Figure 15.2, and can use metal oxides as a catalyst. Aluminum oxide – alumina – is used as an effective catalyst. And while this is an established procedure, research in the area continues [2, 3].

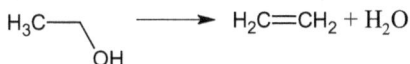

$$H_3C{-}\ \ \underset{OH}{} \longrightarrow H_2C{=}CH_2 + H_2O$$

Figure 15.2: Dehydration of ethanol.

15.3 Catalyst production

15.3.1 Aluminum oxide

Raw aluminum oxide is the main component of bauxite, the mineral from which virtually all aluminum metal is refined. It must have impurities (often, iron-based impurities) removed, as part of its production. Figure 15.3 shows the basic chemistry.

https://doi.org/10.1515/9783110542868-015

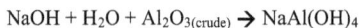

$NaOH + H_2O + Al_2O_{3(crude)} \rightarrow NaAl(OH)_4$

Or

$Al(OH)_{3(crude)} + NaOH \rightarrow NaAl(OH)_4$

Subsequent heating

$NaAl(OH)_4 \rightarrow Al(OH)_3 + NaOH$ separates impurities by solubility

Finally, heating to ca. 1,100 °C

$2\ Al(OH)_3 \rightarrow 3H_2O + Al_2O_3$

Figure 15.3: Alumina production.

Other metal oxides are often produced via direct heating of the reduced metal powder in an oxygen atmosphere.

This is called the Bayer process, after Karl Josef Bayer (also spelled Carl Joseph Bayer), an Austrian chemist who did the initial work in the area and patented his process in the 1890s [4].

As might be expected, most of the alumina produced today is consumed in the production of aluminum metal, via the Hall–Heroult process. However, alumina is still isolated and used as a catalyst.

15.4 Catalyst fate

Although alumina is not particularly toxic, care is taken in the disposal of spent alumina catalyst because of the materials that may have been taken up during its lifetime as a working catalyst. Corporations determine on a case-by-case basis whether or not it is economically feasible to dispose of the old catalyst, or if it should be reclaimed in some fashion.

References

[1] American Chemistry Council. Website. (Accessed 12 October 2020, as: www.americanchemis try.com).

[2] Chemical Processing, More Efficient Ethylene Production Beckons. Website. (Accessed 7 October 2020, as: chemicalprocessing.com/articles/2020/more-efficient-ethylene-production-beckons).

[3] Ethanol to ethylene (B1) – Process design. Website. (Accessed 13 Oct 2020, as: processdesign.mccormick.northwester.edu).

[4] Karl Joseph Bayer. Process of Making Alumina. Patent. US515895A, 1894.

Chapter 16
Ethylene dichloride

16.1 Introduction

Ethylene dichloride has both an established history, being first produced in 1794 in Holland – and quite a variety of names. It has at various times been called ethylene chloride, DCE, 1,2-dichloroethane, 1,2-DCA, Freon-150, and even Dutch liquid (this last one is the earliest name, and is definitely the least systematic!). The major use of this colorless liquid is the production of ethylene chloride, shown in Figure 16.1.

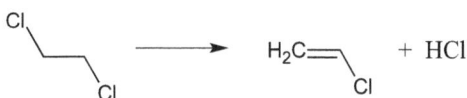

Figure 16.1: Ethylene dichloride use.

Ethylene dichloride is in turn used for the production of polyvinylchloride, or PVC, which can be produced in three different tacticities. This is a plastic produced in such large amounts that it has its own resin identification code (RIC), RIC-3, shown in Figure 16.2, using three of its most well-known symbols [1].

Figure 16.2: Resin identification code for PVC.

In this chapter, it is the production of ethylene dichloride that we will focus upon, and not the polymer end products.

16.2 Reaction chemistry

The direct addition of ethylene to elemental chlorine is the means by which the majority of ethylene dichloride is produced today, an exothermic reaction shown schematically in Figure 16.3. The sources for the reactants are generally crude oil for the ethylene, although there are bio-based sources as well, and the chlor-alkali process for the large-scale production of elemental chlorine.

https://doi.org/10.1515/9783110542868-016

$$H_2C{=}CH_2 + Cl_2 \longrightarrow$$

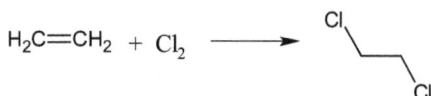

Figure 16.3: Direct addition of ethylene to chlorine.

Iron (III) chloride is the catalyst in this production, as chlorination does not proceed to any meaningful extent – or economically feasible extent – without it.

A second synthetic strategy is the addition of HCl and elemental oxygen to ethylene, referred to as an oxychlorination, shown in Figure 16.4. The by-product here is liquid water, which must be separated from the desired product.

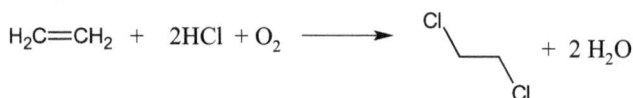

$$H_2C{=}CH_2 + 2HCl + O_2 \longrightarrow \qquad + 2\,H_2O$$

Figure 16.4: Ethylene dichloride production via oxychlorination.

Copper (II) chloride is used in this synthetic method. Of note is that both methods rely on chloride to produce the desired, chlorinated product.

16.3 Catalyst production

16.3.1 Iron (III) chloride – FeCl$_3$, aka ferric chloride

Iron (III) chloride, a bright yellow salt, is generally produced by the direct addition of the elements, but on an industrial scale can be produced as a solution, which is then dried for further use. The reaction chemistry is shown in Figure 16.5.

$8\,HCl_{(aq)} + Fe_3O_{4(s)} \rightarrow FeCl_{2(aq)}$

then

$Cl_{2(g)} + 2\,FeCl_{2(aq)} \rightarrow 2\,FeCl_{3(aq)}$

Figure 16.5: Iron (III) chloride production.

Figure 16.6 shows the basic reaction chemistry of a second important method whereby iron (III) chloride can be produced.

$O_{2(g)} + 4\,HCl_{(aq)} + 4\,FeCl_{2(aq)} \rightarrow 4\,FeCl_{3(aq)} + 2\,H_2O_{(l)}$

Figure 16.6: Iron (III) chloride, alternate production.

In either case, the water must be separated. While companies tend to keep this final step proprietary, it is known that the use of thionyl chloride is one means by which the separation takes place. Interestingly, a direct heating does not drive off water. Rather, it decomposes the iron compound.

16.3.2 Copper (II) chloride – CuCl$_2$

Copper (II) chloride is a dark brown solid when anhydrous, and a bright blue one when it exists as its dihydrate. The anhydrous salt is somewhat deliquescent and will absorb water from the air, albeit slowly.

The production of copper (II) chloride can be through the direct addition of the two elements at high heat (at least 300 °C). The result is a molten salt, as shown in Figure 16.7.

$$Cl_{2(g)} + Cu_{(s)} \rightarrow CuCl_{2(l)}$$

Figure 16.7: Copper (II) chloride production.

Another method of production begins with copper (II) oxide, which is added to ammonium chloride, at similar temperatures to that for the direct addition, as shown in Figure 16.8.

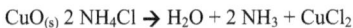

$$CuO_{(s)} \ 2\ NH_4Cl \rightarrow H_2O + 2\ NH_3 + CuCl_2$$

Figure 16.8: Production of CuCl$_2$ from CuO.

Note that in this reaction, the by-products are both water and ammonia, which must be separated to produce the pure salt.

Purification is very often through crystallization of the product.

As well as for the production of ethylene dichloride, chemists are often aware that copper (II) chloride is used on a large scale in what is called the Wacker process, the production of acetaldehyde. Here it functions as a cocatalyst, with palladium (II) chloride, reoxidizing the palladium salt to PdCl$_2$.

16.4 Catalyst fate

Neither catalyst contains a precious metal or a platinum group metal. Their disposal or recovery then is not usually a matter of economic concern. Local and regional laws do dictate how inorganic materials are to be disposed of, to minimize area pollution, meaning that disposal differs in different areas [2].

References

[1]   American Chemistry Council, Plastics Division. Website. (Accessed 13 September 2020, as: plastics.americanchemistry.com).

[2]   United States Geological Survey, Mineral Commodity Summaries. Downloadable at: https://www.usgs.gov, as pubs.er.usgs.gov.

Chapter 17
Ethylene glycol

17.1 Introduction

Ethylene glycol, the smallest molecular weight polyol, is a commodity chemical produced on an industrial scale which has a large suite of sues, although two currently dominate the market. Its Lewis structure is shown in Figure 17.1.

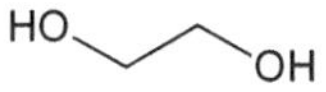

Figure 17.1: Ethylene glycol.

Ethylene, discussed in Chapter 15, is the feedstock for ethylene glycol production. The two biggest uses of ethylene glycol tend to be as a reagent in the production of polyesters and as an antifreeze. But there are other uses as well, and several of which are shown in Table 17.1.

Table 17.1: Ethylene glycol uses.

Purpose or use	Common name	Comment
Polyester production	PETE or PET	Also requires terephthalic acid or terephthalate [1, 2]
Coolant	Antifreeze	Used because of its low freezing point
Drying agent, desiccant		Used because it is hygroscopic
Ink component		As a solvent, increases viscosity
Explosives	EGDN, ethylene glycol dinitrate	Used in some dynamite formulations
Preservative		Used to preserve woods by preventing fungi formation and growth

Curiously, the production of ethylene glycol was brought up to a large scale just over a century ago, when it was used almost exclusively as a component in the production of dynamite. The rise of the plastics industry after the Second World War, as well as the spread of the use of automobiles throughout the twentieth century, led to the currently predominant two uses of ethylene glycol.

https://doi.org/10.1515/9783110542868-017

17.2 Reaction chemistry

Ethylene glycol is produced from ethylene via an ethylene oxide intermediate. The reaction chemistry can be illustrated as shown in Figure 17.2. The ethylene oxide intermediate can be isolated and used as well.

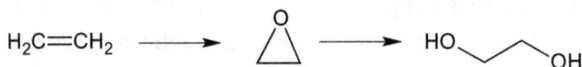

$$H_2C{=}CH_2 \longrightarrow \overset{O}{\triangle} \longrightarrow HO{\diagup}{\diagdown}OH$$

Figure 17.2: Ethylene glycol production.

Water is required to transform the ethylene oxide into the final product but is present as an excess and is not a catalyst in the reaction.

17.3 Catalyst production

To produce the strained ethylene oxide ring, a silver (I) oxide catalyst is required; and the reaction is run in air.

17.3.1 Silver (I) oxide

Figure 17.3 shows the general production of silver (I) oxide. Although sodium hydroxide is shown as the coreactant, most alkali hydroxides can be used to produce it.

$$AgNO_{3(aq)} + NaOH_{(aq)} \rightarrow Ag_2O + NaNO_3 + H_2O$$

Figure 17.3: Silver (I) oxide production.

In turn, silver (I) nitrate is produced by the direct reaction of silver metal with nitric acid. The roots of this reaction stretch back centuries, when it was found that nitric acid would dissolve silver but not gold, thus separating the two precious metals.

   Because of the volume of ethylene glycol produced annually, other catalysts are constantly being examined and tested as well.

17.4 Catalyst fate

Silver is a valuable enough metal that it is tracked by national governments [3, 4]. As well, it is valuable enough that even silver catalyst is recovered for possible reuse, or for subsequent reduction and reclamation of the silver.

References

[1]   PETRA. PET Resin Association. Website. (Accessed 13 October 2020, as: www.petresin.org).
[2]   CPME. Committee of PET Manufacturers in Europe. Website. (Accessed 13 October 2020, as: www.cpme-pet.org).
[3]   U.S.G.S. Mineral Commodity Summaries 2020. Website. (Accessed 13 October 2020, as: pubs.usgs.gov/periodicals/mcs2020/mcs2020.pdf).
[4]   UK Minerals Strategy. Website. (Accessed 13 October 2020, as: www.mineralproducts.org/documents/UK_Minerals_Strategy.pdf).

Chapter 18
Ethylene oxide

18.1 Introduction

As mentioned in Chapter 17, ethylene oxide (EO) can be an intermediate in the production of ethylene glycol, which in turn is routinely used in the production of polyethylene terephthalic ester [1–3]. This is one of its major uses. The Lewis structure of EO is shown in Figure 18.1.

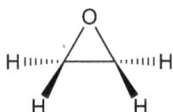

Figure 18.1: Ethylene oxide.

There are other uses of EO beyond the production of ethylene glycol, though. Table 18.1 gives several examples. Note that some are used as a reactant in further chemistry, while others are a direct use of EO.

Table 18.1: Uses of ethylene oxide.

Use	Product	Comments
Antifreeze	Ethylene glycol	Can be used neat or added to blends
Polymer	Polyethylene terephthalic ester	Requires terephthalic acid or terephthalate as well
Pesticide	Can be used directly	Linked to certain cancers
Disinfectant		Used directly in small amounts on foods, surfaces, medical equipment
Polymer	Acrylonitrile	Older method, added HCN to ethylene oxide
Detergents	Ethanolamine	Requires ammonia as well. Large amounts of EO are used in non-ionic surfactants, as well.

18.2 Reaction chemistry

EO is routinely produced from ethylene. Ethylene is the subject of Chapter 15. This means that most EO then uses petroleum as a feedstock, although the past two decades have seen a rise in the use of ethanol – often bioethanol – to produce

https://doi.org/10.1515/9783110542868-018

ethylene through dehydration. The reaction chemistry can be shown simply, as in Figure 18.2.

$$H_2C{=}CH_2 \longrightarrow$$

Figure 18.2: Ethylene oxide production from ethylene.

18.3 Catalyst production

As mentioned in the previous chapter, where EO is an intermediate, a silver oxide catalyst is required to affect the ring closure [4].

18.3.1 Silver (I) oxide

Figure 18.3 shows the simplified reaction chemistry needed to produce silver (I) oxide. Several alkali hydroxides can be used in the production, although sodium hydroxide is the economically most feasible.

$$AgNO_{3(aq)} + NaOH_{(aq)} \rightarrow Ag_2O + NaNO_3 + H_2O$$

Figure 18.3: Silver (I) oxide production.

Again as mentioned in Chapter 17, silver (I) nitrate is itself produced through the direct reaction of nitric acid with silver metal. Nitric acid is enough of an oxidizer that it will dissolve silver metal in ingot form but the reaction proceeds much better when silver is introduced as a powder.

18.4 Catalyst fate

As mentioned in Chapter 17, the value of silver is high enough that this precious metal is not discarded. Rather, it is tracked by national governments as a critical material [5–7]. In 2020, the price of gold metal briefly rose to above US $2,000 per troy ounce, which also affected the price of silver. Since this surge brought the price of silver up, it became even more important to reclaim it, even if it was in an oxidized form, such as Ag_2O.

References

[1]   PETRA. PET Resin Association. Website. (Accessed 13 October 2020, as: www.petresin.org).
[2]   CPME. Committee of PET Manufacturers in Europe. Website. (Accessed 13 October 2020, as: www.cpme-pet.org).
[3]   American Chemistry Council. Website. (Accessed 13 October 2020, as: americanchemistry. com/Ethylene-Oxide/).
[4]   Ozbek, M.O., Onal, I., and van Santen, R.A. Why silver is the unique catalyst for ethylene epoxidation. Journal of Catalysis, 284, 2, (2011), 230–235.
[5]   U.S.G.S. Mineral Commodity Summaries 2020. Website. (Accessed 13 October 2020, as: pubs.usgs.gov/periodicals/mcs2020/mcs2020.pdf).
[6]   UK Minerals Strategy. Website. (Accessed 13 October 2020, as: www.mineralproducts.org/ documents/UK_Minerals_Strategy.pdf).
[7]   Minerals and Mining Publications. Website. (Accessed 13 October 2020, as: nrcan.gc.ca/ maps-tools-publications/publications/minerals-mining-publications/18733).

Chapter 19
Formaldehyde

19.1 Introduction

The smallest molecular weight aldehyde, formaldehyde, is produced on a scale of tens of millions of tons each year. Little of it is sold as is in terms of some user end product, though. Rather, there are numerous products made from the further reaction of formaldehyde and some other organic starting material, and the end product or products are often a variety of resins. The uses for such resins are even more numerous than the variety of these materials. The most common is the urea–formaldehyde resin, which is used extensively in wood composites, as well as fiber board and particle board. The basic scheme of its synthesis is shown in Figure 19.1. It should be noted that while the formaldehyde is shown in the figure as a stand-alone chemical, it is not used as a neat material but rather as some type of solution.

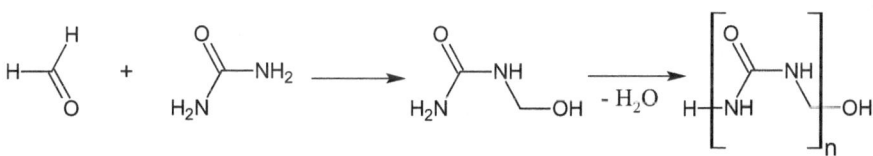

Figure 19.1: Formaldehyde–urea synthesis.

The first reaction is actually an equilibrium but can be made to favor the formation of the precursor to the polymer.

Retrosynthetically, formaldehyde is made from methanol, and this in turn is made from methane, the predominant component of natural gas. This is shown in Figure 19.2. Thus, the ultimate source for formaldehyde is petrochemical, natural gas.

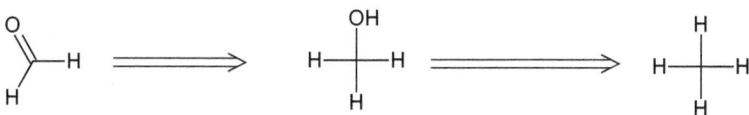

Figure 19.2: Source of formaldehyde.

The Formox™ process is Johnson Matthey's method of producing this from methanol, using iron oxide and molybdenum and/or vanadium in catalytic roles [1]. Other methods and other catalysts have been used as well, most notably crystalline silver.

https://doi.org/10.1515/9783110542868-019

19.2 Reaction chemistry

From methanol, the formation of formaldehyde is represented as shown in Figure 19.3. Perhaps obviously, oxygen must be added in a controlled manner.

$$CH_3OH + \tfrac{1}{2}O_2 \rightarrow H_2CO + H_2O$$

or

$$CH_3OH \rightarrow H_2CO + H_2$$

Figure 19.3: Formaldehyde production from methanol.

The second form of the reaction is simply a dehydrogenation. Interestingly, since the reaction is an endothermic one, some source of heat is required. That heat is often provided by the burning of the by-product hydrogen, after it has been isolated.

19.3 Catalyst production

Since there are multiple means by which formaldehyde is produced, the catalysts used in each system are discussed separately.

19.3.1 Silver

The United States Geological Survey tracks over 80 mineral commodities, including silver [2], and indicates that in 2018 the United States imported slightly more than 60% of the silver that it consumed. Major nations from which it is imported include Mexico, Canada, Peru, and Poland. Major domestic producers are in the western states of the United States.

The production of silver in the United States has a long history and is intimately connected with the expansion into the West. Arguably the most famous silver found in the United States is the Comstock Lode, in what is now Nevada, which occurred hundreds of years after several major silver finds in what was first colonial New Spain, then modern Mexico. At that time of the discovery of the Comstock, so much silver was found that it altered the economy of the United States, and ultimately of the world, resulting in the massive production of silver dollars for domestic use in the United States, and of a trade dollar, made for export and use in international trade into what was then the Empire of China. Even today, silver is marketed by numerous nations as one-ounce bullion coins, and examples of which are shown in Figure 19.4. These are inexpensive ways for the public to store some value in a precious metal. But these are not usually suitable forms for silver as a catalyst, since their surface area is low.

Figure 19.4: Canadian Silver Maple Leaf, one-ounce bullion coins.

For the production of formaldehyde, silver crystals must be grown from solution, maximizing the surface area of the metal and the purity of the catalyst. The production is in an electrolytic solution of silver (I) nitrate, a basic diagram of which is shown in Figures 19.5 and 19.6, called a Balbach–Thum cell and a Moebius cell, respectively. Once made, the lifetime of the catalyst is routinely 1 month to 1 year [3], depending on the conditions of each run.

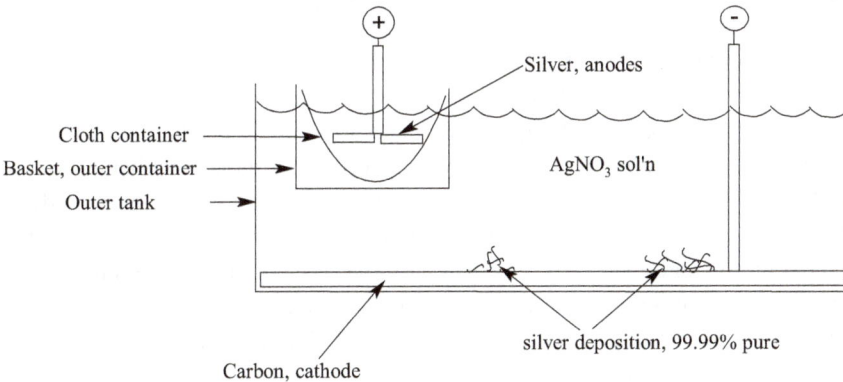

Figure 19.5: Growth of silver for formaldehyde catalyst, Balbach–Thum cell [5].

Note that in the growth process, other metals that are naturally occurring in what can be called the silver source are effectively separated in either type of cell. This small amount of residue may contain other precious metals which are worth recovering, and which must always be kept away from the purified silver. As well, note

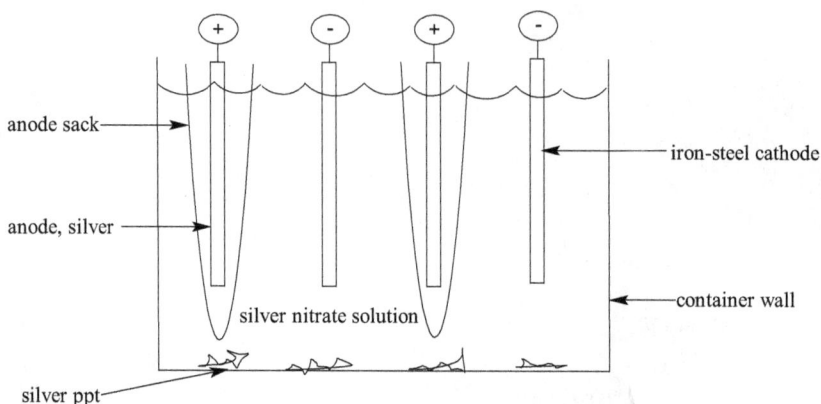

Figure 19.6: Growth of silver for formaldehyde catalyst, Moebius cell [5].

that the Moebius cell has electrodes positioned vertically, which results in set ups that require less space than a Balbach-Thum apparatus.

19.3.2 Iron oxide with molybdenum and/or vanadium

The use of iron oxide with molybdenum or vanadium is called the Formox process and is a relatively recent one [4]. A wide variety of iron oxides are found naturally, and thus iron oxide for use as a catalyst is not particularly difficult to procure.

Iron oxide is not tracked by the U.S.G.S. *Mineral Commodities Summary* [2], simply because iron and steel are tracked. The use of iron oxide as a catalyst is a minor use of any form of iron ore compared to the use of it for the refining of elemental iron.

Molybdenum is produced in large enough quantities domestically that the United States has been a net exporter for several years. Several US mines produce molybdenum as a by-product of copper production, while three extract it as a primary product. Roughly, three quarters of molybdenum use is for the production of what are called superalloys or iron and steel. Thus, its use as a catalyst is a minor overall application.

Vanadium is not mined as a primary product in the United States but rather is imported from South Korea, Canada, Austria, and the Czech Republic.

19.4 Catalyst fate

Silver is an expensive enough metal that it is almost always recycled, including from its use as a catalyst. In the Formox process, the valuable metals that are generally tracked for recycling and reuse are molybdenum and vanadium [2].

References

[1] Johnson Matthey. Website. (Accessed 11 December 2018, as: https://matthey.com/markets/chemicals/formaldehyde).
[2] U.S. Department of the Interior, U.S. Geological Survey, Mineral Commodity Summaries 2018. Downloadable, at: https://minerals.usgs.gov/minerals/pubs/mcs/2018/mcs2018.pdf.
[3] Millar, G.J., and Collins, M. Industrial production of formaldehyde using polycrystalline silver catalyst. Industrial & Engineering Chemistry Research, 56, (2017), 9247–9265.
[4] Formox. Website (Accessed 6 March 2019 as: www.formox.com)
[5] EP0775763B1. Silver electrolysis method in Moebius cells, 1995.

Chapter 20
Hydrogen

20.1 Introduction

Hydrogen, the lightest element on the periodic table, never occurs on the Earth in its reduced state. A great deal of it is bound up in water but it occurs in other compounds as well – notably those compounds in crude oil and natural gas. The splitting of water to its component elements, as shown in Figure 20.1, can be accomplished via an electrochemical process but is not yet economically feasible on a large scale. Accomplishing this direct splitting in some economically feasible manner continues to be one of the grand challenges of chemistry for the near future.

$$2\ H_2O_{(l)} \rightarrow 2\ H_{2(g)} + O_{2(g)}$$

Figure 20.1: Electrochemical splitting of water.

This reaction remains of interest today, simply because if this direct division of water into its elements can be made profitable, as mentioned, it would be a means of producing hydrogen that uses a virtually unlimited feedstock – water – than what is currently used – natural gas.

20.2 Reaction chemistry

What is called the steam reforming of some hydrocarbon – almost always natural gas – is an economically advantageous means of producing elemental hydrogen gas. Figure 20.2 shows the basic reaction chemistry. The products are called "syn gas" for synthesis gas.

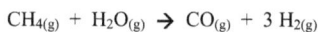

$$CH_{4(g)} + H_2O_{(g)} \rightarrow CO_{(g)} + 3\ H_{2(g)}$$

Figure 20.2: Steam reforming to produce hydrogen.

What the figure does not show is the temperature required (from 1,000 to 1,400 K) as well as the pressure required (roughly 20 atm). It also does not show that for different starting batches of natural gas, for which methane is the main but not sole component, different amounts of carbon monoxide and hydrogen are formed. Also, depending on the feed, coke can be formed as a by-product.

https://doi.org/10.1515/9783110542868-020

Of interest to us here, Figure 20.2 does not indicate the use of nickel as a catalyst. Despite these conditions, steam reforming is by far the most common means of producing hydrogen. Industrially, over 90% of hydrogen is produced in this manner.

This method, while effective, remains expensive in terms of its energy input.

Hydrogen is also produced by the electrolysis of water. But that reaction, shown in Figure 20.3, is used to produce sodium hydroxide and chlorine as the main products. Chlorine is discussed in Chapter 11.

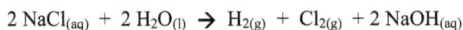

$$2\,NaCl_{(aq)} + 2\,H_2O_{(l)} \rightarrow H_{2(g)} + Cl_{2(g)} + 2\,NaOH_{(aq)}$$

Figure 20.3: Hydrogen production from salt water electrolysis.

This process, called the chlor-alkali process, does not require a catalyst, merely a dedicated source of electricity.

20.3 Catalyst production

Elemental nickel metal is often supported, usually on silica. This allows maximum surface area of the nickel and thus optimizes its catalytic ability [1, 2].

20.4 Catalyst fate

While nickel is not considered a precious metal or an expensive metal because it functions as a heterogeneous catalyst, its recovery is not particularly difficult. In operations that use either platinum or a platinum–palladium mixture, the value of the catalyst is such that it is always recovered when possible [3]. As well, environmental regulations in some areas place a burden on the disposal of metals such as nickel [4].

References

[1]   United Technologies Corporation. "Iron oxide catalyst for steam reforming," U.S. Patent 4,451,578.
[2]   Pieterse, J.A.Z., and van den Brink, R.W. On the potential of nickel catalysts for steam reforming in membrane reactors. Catalysis Today, 156, 3–4, (2010), 153–164.
[3]   United States Geological Survey, Mineral Commodity Summaries, 2019. Downloadable as: mcs2019_all.pdf.
[4]   Sehested, Jens. Four challenges for nickel steam-reforming catalysts. Catalysis Today, 111, 1–2, (2006), 103–110.

Chapter 21
Hydrogen peroxide

21.1 Introduction

Hydrogen peroxide finds a large number of uses, such as bleaching in paper production, or as a component of detergents, although the one with which most consumers are familiar with is the disinfection of minor cuts and scrapes (with a 1% solution concentration). The molecule is a relatively unstable one, with the oxygen atoms within it easily reducing from their formal −1 state to some oxide with a −2 charge, often water. Figure 21.1 shows the Lewis structure of hydrogen peroxide.

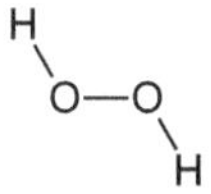

Figure 21.1: Lewis structure of hydrogen peroxide.

The term "unstable" does not mean though that hydrogen peroxide cannot be produced and transported in large quantities. Figure 21.2 shows a tanker truck of hydrogen peroxide en route on one of the US interstate highways.

Figure 21.2: Hydrogen peroxide transport.

https://doi.org/10.1515/9783110542868-021

21.2 Reaction chemistry

The current production of hydrogen peroxide does not directly combine the two elements, as any such attempts to do so ultimately result in the reduction of oxygen to a formal −2 oxidation state, and the production of water as a major product. Rather, today virtually all hydrogen peroxide is produced through what is known as the anthraquinone process. The process was patented just prior to the Second World War, 1939, and can use several different versions or derivatives of anthraquinone. It has also undergone several improvements over the course of years [1–4]. Figure 21.3 shows the structure of ethyl anthraquinone, which has proven to be a very effective catalyst for this transformation.

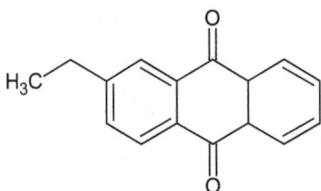

Figure 21.3: Ethylanthraquinone.

Other alkyl derivatives of anthraquinone can be utilized as well but the ethyl derivative has found the most use industrially, as seen through numerous trials. The catalytic cycle can be shown simply, as in Figure 21.4.

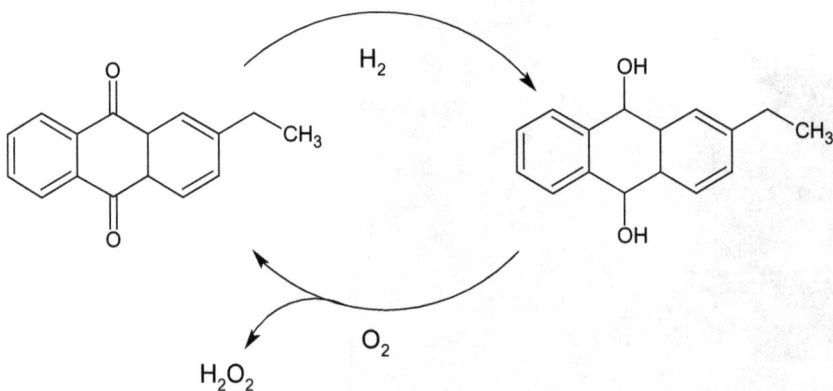

Figure 21.4: Hydrogen peroxide production.

Note that in Figure 21.4, anthraquinone undergoing an oxidation–reduction to anthrahydroquinone is shown. This occurs through the use of a palladium catalyst in the presence of elemental hydrogen. The anthraquinone is then regenerated through an auto-oxidation. This produces hydrogen peroxide as what can be considered a by-product, although this is the desired, commercially viable target material. The oxygen source for the hydrogen peroxide is usually compressed air, which has sufficient elemental oxygen for the reaction. The only other step to the process is the removal of the hydrogen peroxide from solution.

The original patent for this process dates back, as mentioned, to the late 1930s. But improvements involved the trial of several different metals as part of the catalyst cycle. As quoted in a later patent:

> As catalyst, there is employed in the present invention, any of the suitable catalysts known to foster the reduction of the quinone group to the hydroquinone group, as for instance Raney nickel or one of the noble metals ruthenium, rubidium, platinum, rhodium, or palladium. Palladium being one of the commercially better known catalysts for use in the anthroquinone process, is illustrated herein as the catalyst employed, and when so employed is present upon a catalyst carrier and in a fixed bed . . .. [3]

This is noteworthy because several other metals have been tried successfully. Those currently used are simply those that are economically feasible to purchase in the needed quantities that still ensure sufficient H_2O_2 production.

21.3 Catalyst production

Because this process is widely used, there have been several different means by which the proper anthraquinone derivative is developed. It may seem logical to start with anthracene and selectively oxidize and de-aromatize the central ring, as shown in Figure 21.5, but this does not have to be the case.

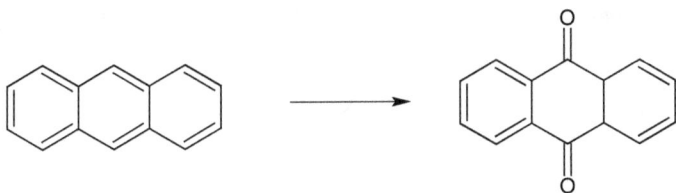

Figure 21.5: Anthraquinone synthesis.

Recently, BASF has developed a technique by which 1,3-diphenylbutene is reacted to affect a ring closure and oxidation, as shown in simple form in Figure 21.6.

The production of metal catalysts such as palladium has been discussed in other chapters. The key aspect of such production is, as in almost all cases, maximizing the surface area of the catalyst.

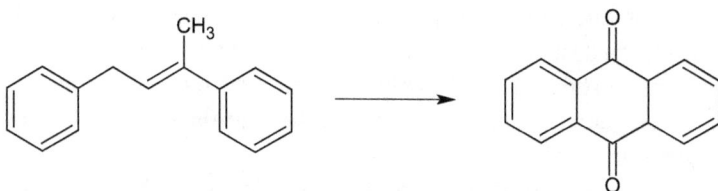

Figure 21.6: Anthraquinone synthesis from 1,3-diphenylbutene.

21.4 Catalyst fate

The production of anthraquinone is the most expensive aspect of the process, at least initially when such a process is being set up. Therefore, recovery and reuse is important. Whenever possible, anthraquinone is recycled and reused.

The recovery of palladium or any other precious metal is one of extracting the material from solution, usually not a particularly difficult task, since it is the separation of a liquid from a solid.

References

[1] Production of hydrogen peroxide. Hans-Joachim Rledl and Georg Pfleiderer, U.S. Patent 2,158,525.

[2] Anthraquinone Process. Colie Lawrence Jenkins, Fred Bronson Kirby, and Theodore Augur Koch. U.S. Patent 5,772,977A.

[3] Production of hydrogen peroxide by anthroquinone process in the presence of a fixed bed catalyst. Donald H. Porter, FMC Corporation, US Patent 3,009,782.

[4] Production of Hydrogen Peroxide by Anthraquinone Process, Nathan D. Lee, U.S. Patent 3,615,207.

Chapter 22
Isopropanol

22.1 Introduction

Isopropanol, also called 2-propanol or rubbing alcohol, is one of the most widely used industrial solvents. Its Lewis structure is shown in Figure 22.1. The general public tends to know of it or think of it through the sale of relatively small bottles of it as rubbing alcohol, a disinfectant. The history of large-scale production goes back to 1920 and the Standard Oil Company. Production at that time was directed more toward making war material than it is today, especially the production of cordite, a smokeless explosive.

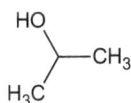

Figure 22.1: Isopropyl alcohol.

Currently, production of isopropanol is large scale because there are and continue to be numerous uses for it. Shell Chemical states at its website that:

> Isopropyl Alcohol is used in a variety of applications including as a solvent for industrial processes and coating; as a component in cleaning, car care and deicing products; as a wetting agent for printing inks and as a feedstock in the manufacture of ester and Mogas/Luboil additives.   [1]

Other large suppliers, such as Exxon Mobil and Dow, indicate that they provide isopropanol for these and other uses [2, 3]. Again, it is an industrial solvent with a wide profile of uses.

22.2 Reaction chemistry

Traditionally, the starting feedstock for isopropanol production has been propylene. As shown in Figure 22.2, it has been found that reacting it with sulfuric acid, which can be recovered, produces isopropanol.

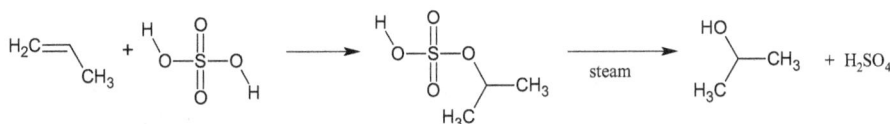

Figure 22.2: Traditional isopropanol production.

https://doi.org/10.1515/9783110542868-022

The more recent method avoids the use of sulfuric acid and does eventually re-quire a fixed-bed catalyst that is phosphoric acid based. The basic reaction chemis-try is shown in Figure 22.3. Note that propylene is still the hydrocarbon feedstock.

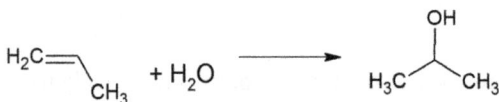

Figure 22.3: Isopropanol production using phosphoric acid catalyst.

This direct addition can also be carried out at a lower temperature than that which was required in previous processes, as a liquid-phase reaction but then requires a tungsten catalyst that is soluble in that phase. As well, this process requires a purer grade of propylene than that which uses sulfuric acid.

22.3 Catalyst production

22.3.1 Phosphoric acid

Phosphoric acid is produced in large quantities, for use as a fertilizer. It can be pro-duced as shown in Figure 22.4. Its use as a catalyst, while important, is secondary to its use as a fertilizer. As a catalyst though, its purity has to be higher than that used as fertilizer.

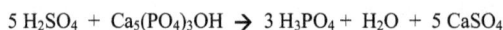

$$5\,H_2SO_4 + Ca_5(PO_4)_3OH \rightarrow 3\,H_3PO_4 + H_2O + 5\,CaSO_4$$

or

$$Ca_5(PO_4)_3F + H_2SO_4 \rightarrow H_3PO_4 + Na_2SiF_6 + CaSO_4$$

Figure 22.4: Phosphoric acid production.

22.3.2 Tungsten catalyst

Improvements in catalysts, in terms of expected turnover, surface area, and life span, continue in many processes, including the production of isopropanol. This is worth mentioning because the original tungsten-based catalyst for this is: "Hydration of propylene with blue oxide of tungsten catalyst," dated July 1954 [4]. This rather color-ful name for the catalyst (pun intended?), blue oxide of tungsten, was believed to be W_2O_5, which does indeed have a blue color. Subsequent research into the material indicates it as $W_{18}O_{49}$, which can be formed by reacting tungsten metal with WO_3 at temperatures of approximately 700 °C.

22.3.3 Raney nickel

And finally, when starting with acetone, Raney nickel, palladium, and ruthenium can be used to affect the hydrogenation of acetone to isopropanol. The production of Raney nickel has been discussed in Chapter 5.

22.4 Catalyst fate

Phosphoric acid is, as mentioned, a large-scale product in its own right. There is little need or economic incentive to recover and reuse this material.

The cost of palladium and ruthenium is high enough that catalyst recovery is common when isopropanol is manufactured. Likewise, the recovery of catalysts containing tungsten or Raney nickel is usually advantageous from an economic stance, and may be mandated by different local or regional laws.

References

[1] Shell Chemical. Website. (Accessed 3 December 2019, as: https://www.shell.com/business-customers/chemicals/safe-product-handling-and-transportation/product-stewardship-summaries/_jcr_content/par/expandablelist/expandablesection_1177843978.stream/1526084629217/2df2262cb897b24b2b1b9ffc7fe19ebaeb76b9fa/isopropyl-alcohol-pss-december-2017.pdf).
[2] ExxonMobil Chemical Company. Website. (Accessed 3 December 2019, as: https://www.exxonmobilchemical.com/en/exxonmobil-chemical).
[3] Dow. Website. (Accessed 3 December 2019, as: https://www.dow.com/en-us/document-viewer.html?ramdomVar=4546470258120207432&docPath=/content/dam/dcc/documents/en-us/productdatasheet/327/327-00031-01-isopropanol-tds.pdf).
[4] Preparation of isopropanol. European Patent Application. Application number: 90300403.4.5. US Patent 2,683,753. Hydration of propylene with blue oxide of tungsten catalyst.

Chapter 23
Linear alpha olefins

23.1 Introduction

Linear alpha olefins (LAOs) or normal alpha olefins are a group of long molecules which all share a common feature, a double bond at the terminal, or alpha, position. The uses for them are varied, and generally depend upon the length of the aliphatic chain. In theory, however, all LAOs can be used to adjust the density of polyethylene, by being added to the reaction in controlled amounts as the polymerization occurs.

Table 23.1 shows a range of LAOs, based on molecular weight, and their primary uses. Note that the molecules shown are in progressively larger units by two carbon atoms. This is because such molecules are routinely made using ethylene as a starting material, and thus they grow or extend two carbon atoms at a time.

Table 23.1: LAOs and their main uses.

Name	Formula	Structure	Major uses
1-Butene	C_4H_8	$H_2C=CHC_2H_5$	PE production, aldehyde production
1-Hexene	C_6H_{12}	$H_2C=CHC_4H_9$	PE production, aldehyde production
1-Octene	C_8H_{16}	$H_2C=CHC_6H_{13}$	PE production, aldehyde production
1-Decene	$C_{10}H_{20}$	$H_2C=CHC_8H_{17}$	Polyalpha-olefin synthetic lubricant
1-Dodecene	$C_{12}H_{24}$	$H_2C=CHC_{10}H_{21}$	Detergents and surfactants
1-Tetradecene	$C_{14}H_{28}$	$H_2C=CHC_{12}H_{25}$	Detergents and surfactants
1-Hexadecene	$C_{16}H_{32}$	$H_2C=CHC_{14}H_{29}$	Lubricating fluid, surfactant, paper sizing
1-Octadecene	$C_{18}H_{36}$	$H_2C=CHC_{16}H_{33}$	Lubricating fluid, surfactant, paper sizing
C_{20}–C_{24} blend	$C_{20}H_{40}$ minimum	Mix	Heavy linear alkyl benzene production
C_{24}–C_{30} blend	$C_{24}H_{48}$ minimum	Mix	Heavy linear alkyl benzene production
C_{20}–C_{30} blend	$C_{20}H_{40}$ minimum	Mix	Heavy linear alkyl benzene production

23.2 Reaction chemistry

Several large corporations produce LAOs, through somewhat different processes. In alphabetical order, the following processes are currently in use:

https://doi.org/10.1515/9783110542868-023

1. Ethyl Corporation – the Ineos process
2. Gulf, Chevron Phillips process
3. Idemitsu petrochemical process
4. IFP dimerization – used specifically to produce 1-butene
5. Phillips ethylene trimerization process – used to produce 1-hexene
6. SABIC–Linde Alpha-Sablin process
7. Shell Oil Company Process, aka SHOP

While company names are sometimes the same or very close to the name of the process, some differ because one company has been acquired by another, or has for some other reason changed its name [1–8].

23.3 Catalyst production

Because there are several different major producers of LAOs, it is difficult to find open literature that discusses what catalysts are used for their production, and how the catalysts are themselves produced. The information is often proprietary, so that companies are unable to utilize the techniques of a rival. But the following is known from open literature.

23.3.1 The Ineos process

Triethyl aluminum is used in the production of LAOs through this process (Ineos is an abbreviation for INspec Ethylene Oxide Specialties) but not in a traditionally catalytic role, meaning in very small amounts. Nevertheless, the starting material is ethylene, the product is a multiple of ethylene, triethyl aluminum is required and is not part of the product. Figure 23.1 shows the basic Lewis structure of triethylaluminum. Note that it exists as a dimer.

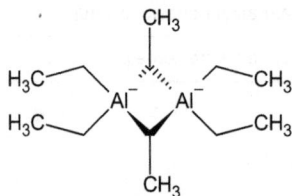

Figure 23.1: Triethylaluminum, as a dimer.

Triethylaluminum in turn can be made from aluminum powder, hydrogen gas, and ethylene. This then becomes an interesting use for ethylene, the production of a catalytic material that is in turn used to transform ethylene into LAOs.

23.3.2 The Gulf process

In the Gulf process, used by Chevron Phillips, triethyl aluminum is used in a catalytic manner, by mixing directly with ethylene. The catalyst is separated from the product through the use of industrial caustic, such as sodium or potassium hydroxide.

23.3.3 SHOP

The acronym is that for Shell Higher Olefin Process. This process was pioneered by Royal Dutch Shell, and uses a nickel complex, shown in Figure 23.2.

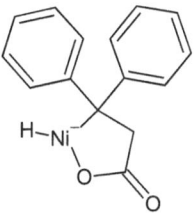

Figure 23.2: SHOP catalyst.

Note that in the catalytic cycle, the hydrogen atom attached to the central nickel atom is displaced by the hydrocarbon that is being lengthened to form the LAO.

23.3.4 Chevron Phillips

A chromium/pyrolle complex is utilized, as shown in Figure 23.3. It has been found to be optimal for the trimerization of ethylene, producing 1-hexene in greater than 99% yield. There is also continuing research on this catalyst to determine if the process can be further improved [9].

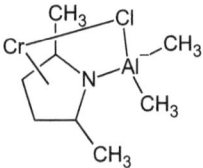

Figure 23.3: Chevron Phillips catalyst.

23.3.5 Idemitsu

In this case, the catalyst that is required can have several possible compositions. According to the US Patent 7393991B2:

. . . a zeolite catalyst and/or montmorillonite catalyst . . .

can be used. Further, this patent states:

> Pt, Ru, Ni etc. supported on alumina, solid acid catalysts such as zeolites (eg. Ferrierite and SAPO) or clay, or their combined metal/solid acid catalysts are known and are already industrially in practice. [10]

This is a further indication of what we have already mentioned that the process of producing LAOs is a widely studied one that is not dependent on a single catalyst.

23.3.6 Exxon research and engineering

In this newer process, alkyl aluminum halides are still required, plus an undisclosed transition metal complex.

23.3.7 DuPont

In this process, once again, an alkyl aluminum is used but as a cocatalyst along with an iron tridentate complex.

23.3.8 IHS

Again, triethyl aluminum is the catalyst used to affect a high-temperature polymerization to higher molecular weight LAOs.

Since alkyl aluminum compounds are used extensively in the production of LAOs, we can use the reaction shown in Figure 23.4 as a representative synthesis because triethyl aluminum is a common aluminum alkyl. In fact there are several synthetic routes to this product but that shown in Figure 23.4 is a very efficient one.

$$2 \text{ Al} + 2 \text{ C}_2\text{H}_{4(g)} + 3 \text{ H}_{2(g)} \rightarrow \text{Al}_2(\text{C}_2\text{H}_5)_6$$

Figure 23.4: Triethyl aluminum production.

Note that the product is written as a dimer. While it functions as a catalyst in its monomeric form, it is routinely produced and stored as the dimer. The material is pyrophoric and must be handled with proper safety protocols.

23.4 Catalyst fate

Since the cost of aluminum alkyls is relatively low – aluminum not being a precious metal – and since the other catalysts mentioned in Section 23.3 also can be low in cost, the catalysts in these reactions tend not to be recovered.

References

[1]   Phillips, C. Website. (Accessed 14 October 2019, as: www.cpchem.com).
[2]   Idemitsu. Website. (Accessed 14 October 2019, as: www.idemitsu.com).
[3]   HIS. Website. (Accessed 14 October 2019, as: www.ihsmarkit.com).
[4]   Ineos. Website. (Accessed 14 October 2019, as: www.ineos.com).
[5]   SABIC-Linde Alpha-Sablin Process. Website. (Accessed 14 October 2019, as:).
[6]   Shell Oil Process, SHOP. Website. (Accessed 14 October 2019, as: www.shell.com).
[7]   Lappin, G.R., and Sauer, J.D. (Eds.). Alpha Olefins Application Handbook. ISBN: 978-0824778958, 1989.
[8]   Singh, T., and Naveen, S.M. A review paper on production of linear Alpha-Olefins by undergoing oligomerization of ethylene. International Journal of Engineering Applied Sciences and Technology, 2, 4, (2017), 83–86.
[9]   Phillips, C., and Naji-Rad, E., et al. Exploring basic components effect on the catalytic efficiency of chevron phillips catalyst in ethylene trimerization. Catalysts, 8, 6, (2018), 224.
[10] Idemitsu: Nubuo Fujikawa. US Patent 7,393,991 B2. Process for producing internal olefin, July 1, 2008.

Chapter 24
Methanol

24.1 Introduction

This is the smallest molecular weight alcohol, and is the one with a large profile of industrial uses. It is often employed as a reaction solvent because it is both nonpolar and has relatively high volatility [1].

24.2 Reaction chemistry

Methanol production begins with carbon monoxide as the carbon source, and elemental hydrogen as the hydrogen source. The reaction, shown in Figure 24.1, appears very simple and direct.

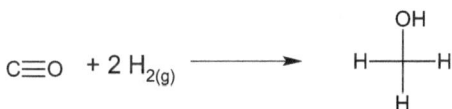

Figure 24.1: Methanol production.

What is not shown is the catalyst that is required, and the general reaction conditions. These are summarized in Table 24.1.

Table 24.1: Conditions for methanol production.

Factor	
Catalyst	Copper/zinc oxide/aluminum oxide/zirconium oxide
	Copper/yttrium or lanthanide/zinc-aluminate support
Pressure	50–100 atm
Temperature	250 °C

From Refs. [2–6].

Clearly, the reaction conditions require the input of a significant amount of energy. This has been a concern since the earliest catalysts developed almost a century ago. One prime example of this is the Cr_2O_3–ZnO catalyst that was used extensively by BASF between the world wars. It required higher temperature and almost three times the pressure as the copper–zinc oxides shown in Table 26.1. It was thus eventually

https://doi.org/10.1515/9783110542868-024

displaced by the latter. Yet even as the catalysts for the process have evolved, attention is paid to maximizing the surface area of the copper [4].

As well, the other metal added with copper is often termed the promoter, simply because of the role it plays [5]. As shown in Table 24.1, there are several elements that can function in this role. The use of one over another can be a matter of proprietary concern, depending upon the facility.

24.3 Catalyst production

The production of alumina has been discussed in previous chapters. Copper metal can simply be ground from some ingot and sieved, so that particles of the proper size are obtained. That with a surface area of approximately 70 m^2/g is found to work well [4].

24.4 Catalyst fate

The production of various catalysts for methanol synthesis, and the cost of the materials, is enough that their recovery is preferred whenever possible. Since these are heterogeneous catalysts, such recovery is feasible in many plants. Reclamation and re-extraction of the metals is often the goal of such processes because the metals are the most expensive component of the catalyst and catalyst system.

References

[1]   U.S. Methanol. Website. (Accessed 13 March 2020, as: usmeoh.com).
[2]   Zhang, Q., Kang, J., and Wang, Y. Development of novel catalysts for Fischer-Tropsch synthesis: tuning product selectivity. ChemCatChem, 2, (2010), 1030–1058.
[3]   Alan Edward Alford Grant. Methanol Production. US Patent 3,950,369, April 13, 1976.
[4]   Takeuchi, M., et al. Methanol synthesis catalyst based on copper and zinc oxide and method for production thereof. EP0864360B1 (accessed 25 May 2020).
[5]   Schoenthal, G.W., and Slaugh, L.H. Methanol synthesis catalyst. January 21, 1986, US4565803A (accessed 25 May 2020).
[6]   Goto, Y., et al. Catalyst for methanol production, method of producing the same and process of methanol production. US9314774B2. April 19, 2016 (accessed 25 May 2020).

Chapter 25
Nitric acid

25.1 Introduction

Nitric acid (HNO_3) has been used as an important industrial chemical for more than a century. It has become one of the world's top commodity chemicals, being produced at the level of several million tons each year, and being used in a wide variety of industries. In the earliest methods of production, a catalyst was not used for the oxidation of nitrogen but rather an electric arc. While this now defunct method was successful – known as the Birkeland–Eyde process – it was expensive enough that the current method – the Ostwald process – rapidly replaced it. The process, named after Wilhelm Ostwald, was patented in 1902 and is the method used in virtually all industrial-scale nitric acid plants.

25.2 Reaction chemistry

The production of nitric acid begins with the production of nitrogen monoxide. The defunct Birkeland–Eyde process did this using an electric arc in elemental nitrogen and elemental oxygen at elevated temperature. The Ostwald process begins with ammonia, as shown in Figure 25.1. This step, an ammonia oxidation, is sometimes called AMOX chemistry.

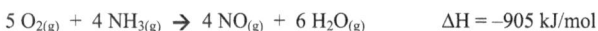

$$5\,O_{2(g)} + 4\,NH_{3(g)} \rightarrow 4\,NO_{(g)} + 6\,H_2O_{(g)} \qquad \Delta H = -905 \text{ kJ/mol}$$

Figure 25.1: Production of nitrogen monoxide from ammonia.

The temperature for this step is approximately 225 °C, and even at this elevated temperature, a rhodium–platinum gauze is required as a catalyst. In most cases currently, platinum alloyed with up to 10% rhodium is used as the catalyst, although early patents on the improvement of the Ostwald process mention several metals that can be used as catalysts:

> [B]y passing over solid or spongy platinum, or a compound of the two, or over metallic iridium, rhodium, or palladium, or over oxides of heavy metals, such as lead, or over silver, copper, iron, chromium, nickel, or copper oxide. The catalytic material is in every case heated to a red heat. [1]

Note that ammonia is also required at the beginning of the production of nitric acid. This means that the Haber process to produce ammonia, discussed in Chapter 6, must be considered an important part of the production of nitric acid.

https://doi.org/10.1515/9783110542868-025

After the production of nitrogen monoxide, the further steps are all continued oxidation of the nitrogen to end at the formal oxidation state of +5. This is shown in Figure 25.2.

$$O_{2(g)} + 2\ NO_{(g)} \rightarrow 2\ NO_{2(g)} \qquad\qquad \Delta H = -114\ kJ/mol$$

$$H_2O_{(l)} + 3\ NO_{2(g)} \rightarrow 2\ HNO_{3(aq)} + NO_{(g)} \qquad\qquad \Delta H = -117\ kJ/mol$$

Figure 25.2: Production of nitric acid from nitrogen monoxide.

As shown in Figure 25.2, nitrogen monoxide is generated as a coproduct. This is re-cycled into the process for reuse in forming nitric acid, the main product. The heat given off in the process (note that all steps are exothermic) is also used and not wasted. Because ammonia is a required starting material, and the Haber process for the pro-duction of ammonia requires heat, it is possible to utilize the heat in this regard.

Note that the nitric acid that is produced must be further concentrated before use. In concentrated form, one of its major uses is the production of fertilizer. An-other is the production of explosives.

Note also that the three starting materials as shown are ammonia, elemental ni-trogen, and elemental oxygen. Since ammonia is produced by the Haber process (as mentioned in Chapter 7), and requires elemental nitrogen and natural gas, one of the starting materials for this is thus natural gas, a fossil fuel.

25.3 Catalyst production

The platinum–rhodium alloys currently in use in the Ostwald process are in the form of a gauze (as noted, platinum–palladium alloy can also be used). This means that the platinum–rhodium alloy must first be extruded as a thin wire, then woven into a gauze. Several sheets of gauze can be stacked, thus maximizing contact with the reac-tants during the process. But the pressure and temperatures are high enough that eventually the alloy begins to separate components.

Manufacturers of the platinum–rhodium catalyst used for nitric acid production generally point out the proprietary nature of their products, but Johnson Matthey states at its website:

> Pgm catalysts for nitric acid production take the form of a gauze made out of fine wire. When nitric acid was first produced commercially in 1904, a platinum-only catalyst was used. Rho-dium was later added for strength and to reduce the amount of platinum lost during conver-sion of the gas. [2]

As with many processes, the heterogeneous catalyst is used as sparingly as possible and with the greatest possible surface area. This is especially true when the catalyst is an expensive material, such as platinum. Johnson Matthey further states at its website:

Palladium-based "catchment" or "getter" gauze was introduced in 1968 to further reduce losses of platinum and rhodium, which can be as high as 300 mg per tonne of acid produced. The catchment sits downstream of the gas flow and collects pgm vapourised from the catalyst.

Until 1990, the catalysts and catchments used in ammonia oxidation were in the form of woven gauze.

Johnson Matthey Noble Metals then introduced a revolutionary knitted gauze, increasing the efficiency of conversion and extending the catalyst life. This has since become the industry standard.   [2]

Of note is the idea that even though the process is mature, further refinements have occurred in the relatively recent past. As well, the statement about the loss of catalyst is important. While 300 mg lost per ton does not seem like a particularly large amount of catalyst, it is worth keeping in mind the sheer tonnage of nitric acid produced annually, roughly 50 million tons. This translates to the loss of 15,000 kg of catalyst, a significant amount of that which is produced annually.

25.4 Catalyst fate

Both platinum and rhodium are rare enough metals [3] that they are recovered for reuse from virtually every application. They are also tracked by the US government and by agencies within other national governments [3–5]. Since the catalytic process here is a heterogeneous one – the platinum–rhodium alloy is in the form of a gauze, several of which can be stacked on each other – catalyst recovery can be performed very efficiently.

References

[1]   Ostwald Process patent, GB 190200698, Ostwald, Wilhelm, "Improvements in the Manufacture of Nitric Acid and Nitrogen Oxides," published January 9, 1902, issued March 20, 1902.
[2]   Johnson Matthey. Website. (Accessed 19 November, 2019, as: platinum.matthey.comAbout PGM/Applications/Industrial/Nitric Acid).
[3]   United States Geological Survey. Mineral Commodity Summaries, 2019. Downloadable as pdf: USGS Mineral Commodity Summaries 2019 report.
[4]   World Mineral Production, 2009 – 2013. British Geological Survey, Natural Environment Research Council, downloadable as: WMP20092013, British Geological Survey.pdf.
[5]   Canadian Mineral Production. Information Bulletin. Website. (Accessed 18 April 2020, as: https://www.nrcan.gc.ca).

Chapter 26
Nylon

26.1 Introduction

The discovery, production, and introduction to the market of the class of materials that is generally called nylon – a series of polyamide molecules – is one of the great stories of a chemical discovery that helped shape the modern world in the twentieth century. The first nylon – which is now called nylon 6,6 – was produced at DuPont by Wallace Carothers in 1935 [1, 2]. In retrospect, the timing of this discovery was perfect, as there was enough time between its initial unveiling and the massive scale up needed to supply the United States Army and Navy with the material that was needed for parachutes, rigging, and other war materials during the Second World War. Seldom has a material gone from initial discovery to industrial-scale production in such a short time.

Today several professional societies exist that promote the use of plastics, including many types of nylon [3–7]. As well, producers of end products made from nylon have trade organizations devoted to the use and sale of such products.

26.1.1 Types of nylon

The numerals in the term "nylon 6,6" refer to the number of atoms in the diamine and in the diacid that are used to produce it. Figure 26.1 shows this repeat structure of this rather simple version of the class of materials. But the aliphatic chain of carbon atoms between the carbonyl carbons and between the amines is areas in which an enormous number of different components can be present, the end result being a wide array of materials that can be called nylons.

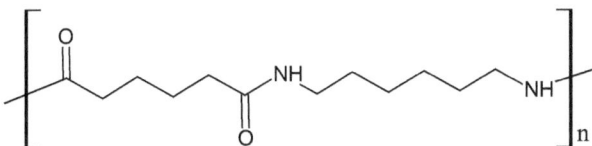

Figure 26.1: Repeat structure of nylon 6,6.

https://doi.org/10.1515/9783110542868-026

26.2 Reaction chemistry

Types of polymerizations

There is more than one set of starting materials that can produce nylon. The use of diamines and diacids is widespread and common, as shown in Figure 26.2. The by-product in such cases is water. But the use of dinitriles has proven effective in some cases, as well. This usually requires some acid catalyst and is called the Ritter reaction, after John J. Ritter who did pioneering work in this area in the late 1940s [8]. Also, the use of acid chlorides in lieu of diacids has proven effective for certain of these polymers.

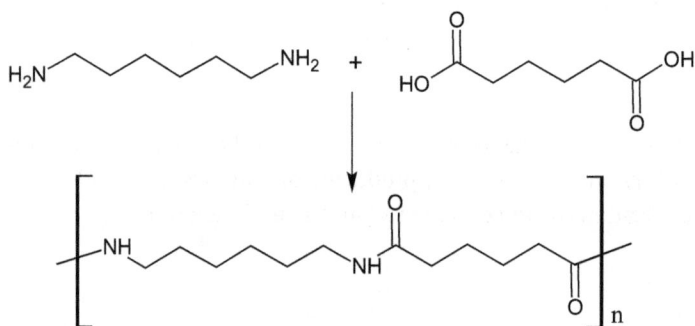

Figure 26.2: Synthesis of nylon 6,6.

26.3 Catalyst production

26.3.1 With dinitriles

Sulfuric acid (H_2SO_4) can be used as an acid catalyst for the production of nylons from di-nitriles. Sulfuric acid is the single most common bulk commodity chemical produced globally, which means it is produced very inexpensively. It is used in the production of fertilizers much more than as a catalyst, although its role as a catalyst is important. Chapter 40 is devoted to the production of it, later in this book.

Phosphoric acid (H_3PO_4) is another acid that has been used to promote nylons production. Like sulfuric acid, it is produced on an enormous scale, and like sulfuric acid, it is often used in the production of fertilizers.

26.3.2 Other acids

Dinitriles can be reacted with water in a stoichiometric amount, also in the presence of a catalytic amount of some dicarboxylic acid, such as adipic acid – the structure of which is shown in Figure 26.3 – to form nylon polymers [9]. This process still does require heat (roughly 1,000 °C) to maximize the reaction yield.

As well, nickel and cobalt catalysts can be used to hydrogenate dinitriles, often on some support. The precise structure of them is often proprietary and kept by the companies that utilize them.

Figure 26.3: Structure of adipic acid.

26.4 Catalyst fate

Sulfuric acid is produced so inexpensively that it is seldom economically feasible to recover it. However, antipollution regulations and laws in some areas have made its recovery a priority.

References

[1]   Wallace Carothers and the Development of Nylon, National Historic Chemical Landmark. Website. (Accessed 19 October 2020 as: acs.org/content/acs/en/education/whatischemistry/landmarks/carotherspolymers.html).
[2]   Hermes, M.E. Enough for One Lifetime: Wallace Carothers, Inventor of Nylon. History of Modern Chemical Sciences, ISBN: 978-0841233317.
[3]   American Chemistry Council. Website. (Accessed 19 October 2020, as: www.americanchemistrycouncil.com/About).
[4]   EC21, Global B2B Marketplace. Website. (Accessed 19 October 2020, as: ec21.com/ed-market/US/nylon.html).
[5]   Plastics Europe. Website. (Accessed 19 October 2020 as: plasticseurope.org/en).
[6]   Chemistry Industry Association of Canada. CIAC. Website. (Accessed 19 October 2020, as: canadianchemistry.ca).
[7]   Chemicals and Petrochemicals Manufacturer's Association, India. CPMAI. Website. (Accessed 19 October 2020, as: www.cpmaindia.com/nfy_about.php).
[8]   Johnson, F., and Madronero, R. Heterocyclic syntheses involving nitrilium salts and nitriles under acidic conditions. Advances in Heterocyclic Chemistry, 6, (1966), 95–146.
[9]   U.S. Patent. Hayes, R.A., Marks, D.N., and van Eijndhoven, M. Process for the hydrolysis of adiponitrile and the production of nylon 6,6 using dicarboxylic acids as the sole catalyst. US6075117A, 1998.

Chapter 27
Phenol

27.1 Introduction

Phenol is one of the simple aromatic compounds that have found its use in many industrial applications, although the largest may be as an antiseptic and a disinfectant. Millions of tons are produced annually. It is a volatile, white solid at room temperature that can easily be crystallized. The material should be handled with care, as contact with moist skin can cause chemical burns. Figure 27.1 shows its Lewis structure.

Major producers of phenol include INEOS – the world's largest – Shell Global, Borealis, Cepsa Quimica, Mitsui Chemicals, Formosa Chemicals, and Kumho P&B Chemicals, although there are several others. Virtually all of these produce cumene as well, an intermediate in the production of phenol [1–4].

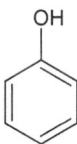

Figure 27.1: Lewis structure of phenol.

27.2 Reaction chemistry

Phenol is used widely, and therefore, as mentioned, several large companies manufacture it. In turn this means that several variations of processes have been developed to produce it. The major method is discussed here. Figure 27.2 shows the starting materials for the production of phenol, and the coproduction of acetone, which always occurs.

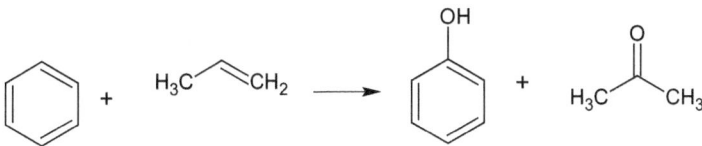

Figure 27.2: Production of phenol and acetone.

The reaction has been studied extensively, and the current means by which the two products are formed is very often called the cumene process. This is because an

https://doi.org/10.1515/9783110542868-027

isolable intermediate, cumene, can be produced. Getting from this to phenol involves a rearrangement that positions an oxygen atom directly on the benzene ring. Because this rearrangement is called the Hock rearrangement, the entire process is sometimes known as the Hock process. This process represents more than 90% of the production of phenol annually.

27.2.1 The cumene process

In this process, benzene and propene (propylene) are the organic feedstocks, and oxygen must be present for the oxidation from benzene to phenol, and for the by-product oxidation of propylene to acetone. Oxygen can be provided from the air. The reaction runs at elevated temperature (ca. 250 °C) and pressure (30 atm), and requires a Lewis acid as a catalyst. Phosphoric acid has a long history in serving as that catalyst, although aluminum halides ($AlCl_3$) function well in this role as well [2–4].

27.2.2 Other methods of phenol production

Besides the cumene process, companies such as Dow Chemical have worked on the production of phenol from chlorobenzene, as well as from toluene, as shown in Figure 27.3. Additionally, the production of phenol from benzene and nitrous oxide shows promise as a greener method of phenol production, as shown in Figure 27.4, since it does not produce any potentially harmful by-products. None of these methods is yet poised to overtake the well-established cumene process.

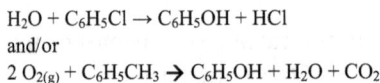

$H_2O + C_6H_5Cl \rightarrow C_6H_5OH + HCl$
and/or
$2\,O_{2(g)} + C_6H_5CH_3 \rightarrow C_6H_5OH + H_2O + CO_2$

Figure 27.3: Alternative methods of phenol production.

$N_2O_{(g)} + C_6H_6 \rightarrow N_{2(g)} + C_6H_5OH$

Figure 27.4: Possible greener method of phenol production.

27.3 Catalyst production

In the earliest years, the catalyst that enabled this reaction was sulfuric acid, H_2SO_4. The production of this is discussed in Chapter 40. More recently, phosphoric acid and aluminum trichloride have been used. As well, a specific form of acidified bentonite clay has been reported as being an effective catalyst.

The large-scale production of phosphoric acid as well as aluminum chloride was discussed in Chapter 12.

Bentonite clays are mostly montmorillonite, a form of phyllosilicate minerals. Formulas vary depending on where they are found and from what source they are extracted. Broadly, the formula can be represented as $(Na,Ca)_{0.33}(Al,Mg)_2(Si_4O_{10})(OH)_2 \cdot nH_2O$. They are mined in numerous parts of the United States and the world; and their use as a catalyst is a minor one compared to ball clay, wall and floor tile, and sanitaryware [5].

27.4 Catalyst fate

Once again, the use and reuse of these catalysts are routinely a matter of economic necessity. All are a very small fraction of the cost for the production of phenol, and thus their disposal is not a problem, unless there is some cost in a region for the disposal of chemical waste.

References

[1]   Tatake, P.A., Kumbhar, P.S., Singh, B., and Fulmer, J.W. Process for producing phenol.
      European Patent. EP1732869A1. (Accessed 1 June 2020).
[2]   Feng, J., et al. Synthesis of bentonite clay-based iron nanocomposite and its use as a
      heterogeneous photo Fenton catalyst. U.S. Patent. US2006/0076299A1, 2006.
      (Accessed 1 June 2020).
[3]   Dandekar, A.B., et al. Process for producing cumene. U.S. Patent. US6888037B2.
      (Accessed 4 June 2020).
[4]   Burattini, M., and Bagatin, R. Process for the preparation of phenol from cumene. U.S. Patent.
      US20120283486A1. (Accessed 4 June 2020).
[5]   United States Geological Survey. Mineral Commodity Summaries, 2019. Downloadable.

Chapter 28
Phosgene

28.1 Introduction

Phosgene continues to be an extremely useful commodity chemical but also remains one intimately associated with the early chemical warfare. This simple molecule, shown in Figure 28.1, was used in the First World War as a means of killing
men in the trenches of the Western Front, either directly through inhalation, since it
is heavier than air, or by forcing them out of the trenches into the path of gunfire.

Figure 28.1: Phosgene.

As a useful chemical, the history of phosgene goes back to the production of some
dyes in the late nineteenth century. Today it is known for its extensive use in the
production of isocyanates, very often methylene diphenyl diisocyanate, and toluene
diisocyanate, both shown in Figure 28.2. These in turn are used in the production of
polyurethanes.

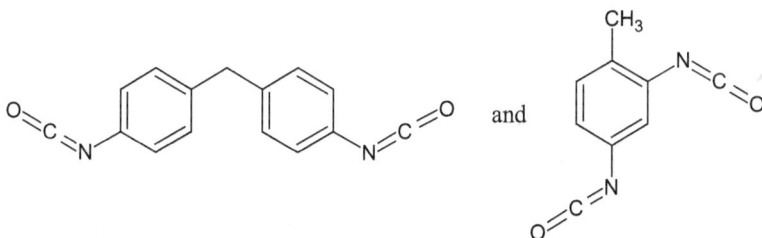

Figure 28.2: Common isocyanates.

Because of the toxic nature of phosgene, it is almost always produced and consumed on site, with extreme safety precautions. This prevents any possibility of a
release during transport, as transport from one site to another is eliminated. Roughly
3 million tons are produced annually, which means eliminating transport is also a
significant economic savings. Despite its use for peaceful purposes, because it is so
easy to make and weaponize, any production of 30 t/year or higher must be reported
to the Organization for the Prohibition of Chemical Weapons.

https://doi.org/10.1515/9783110542868-028

28.2 Reaction chemistry

The reaction that forms phosgene is a straightforward addition reaction, shown in Figure 28.3, between chlorine and carbon monoxide. The reaction requires an elevated temperature of approximately 50–150 °C and should not be run higher than this because of competing side reactions.

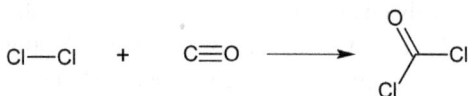

Figure 28.3: Phosgene synthesis.

As well, a catalyst is required for the reaction. Activated carbon has been found to be very effective in optimizing yields [1].

28.3 Catalyst production

Activated carbon – sometimes called activated charcoal – is produced by the controlled, lean combustion of some hydrocarbon source, very often charcoal. It is produced to have the maximum surface area possible, in the form of pores with low volume. Surface areas are often on the order of 3,000 m^2/g.

A wide variety of plant sources can serve for the production of activated carbon, including wood, bamboo, and coconut husks. But coal and pitch can serve as well. It is not easy to draw a reaction showing the chemistry of this production. But the broad steps are as follows:
- pyrolysis in inert atmosphere, ca. 600 °C;
- subsequent exposure to oxygen, again ca. 600 °C.

Lower temperature activation can be achieved by mixing the carbon with an acid such as phosphoric acid (H_3PO_4) or a strong base such as sodium hydroxide (NaOH) or potassium hydroxide (KOH) at temperatures as low as 300 °C. The cost savings involved in this type of method is that of using lower temperatures [2].

28.4 Catalyst fate

Activated carbon is extremely inexpensive compared to many of the other catalysts mentioned in these chapters. It can be discarded if it has not been mixed with other materials.

References

[1]   Riegel, H. Production of phosgene. U.S. Patent. US4346047A. (Accessed 4 June 2020).
[2]   Slyh, J.A., Milton, J., and Doying, E.G. Production of activated carbon. U.S. Patent.
      US2508474A, 1945. (Accessed 4 June 2020).

Chapter 29
Polycarbonate

29.1 Introduction

Polycarbonates represent another class of plastic materials, of carbon-based polymers, that are produced on an enormous scale annually. They tend to be tough, rigid materials, and thus find uses based on their physical properties. The molecular feature common to them all is the $O=C(OC)_2$-unit, as shown in Figure 29.1. The use of polycarbonate, although large scale, has not resulted in these materials being assigned a resin identification code (an RIC) as has been done for plastics like polyethylene terephthalate (RIC 1) or polystyrene (RIC 6). But these plastics are made on a large enough scale that regional and national trade organizations and advocacy groups do exist that concern themselves with both uses and recycling [1–10], and the RIC of 7 is used for polycarbonates.

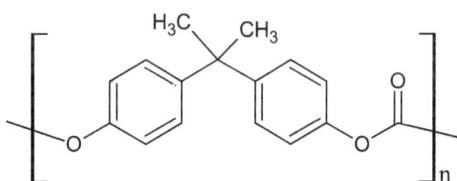

Figure 29.1: Polycarbonate repeat unit.

Several polycarbonates utilize bisphenol A as a starting material, shown in Figure 29.2, which makes for a rigid final product.

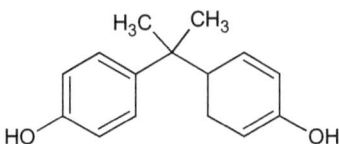

Figure 29.2: Bisphenol A.

In many cases, where bisphenol A is a starting material, the comonomer is phosgene. Because of the poisonous nature of phosgene, efforts continue to find means of production of polycarbonates that do not involve it. This can certainly involve the use of catalysts.

https://doi.org/10.1515/9783110542868-029

29.2 Reaction chemistry

The production of polycarbonate can be shown in a simplified manner, as shown in Figure 29.3, showing the starting diol such as bisphenol A, as well as the di-chloride – the latter usually being phosgene.

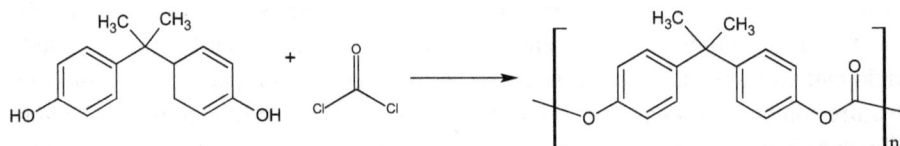

Figure 29.3: Polycarbonate production.

As with other materials, while this does not appear to require a catalyst – and some polycarbonate reactions do not – when one is required, it is not shown in the reaction. In the past, titanium (IV) chloride, aka titanium tetrachloride, has been found to be an effective catalyst, depending upon the two monomers [11]. Also, more recent patented processes have been found, which utilize palladium or cobalt [12].

29.3 Catalyst production

29.3.1 Titanium catalyst

Titanium (IV) chloride (aka titanium tetrachloride) can be produced by the reduction of titanium ores such as ilmenite, as shown in Figure 29.4, with the addition of chlorine gas and has found extensive use as a polymerization catalyst [11–14].

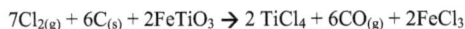

$$7Cl_{2(g)} + 6C_{(s)} + 2FeTiO_3 \rightarrow 2\ TiCl_4 + 6CO_{(g)} + 2FeCl_3$$

Figure 29.4: Production of titanium tetrachloride.

When using some mineral source, such as ilmenite, by-products formed that later must be separated, since ore sources are typically not pure substances. Thus, it is often useful to begin with refined TiO_2 when producing titanium tetrachloride.

29.3.2 Zinc glutarate

Produced as a catalyst for different reactions, zinc glutarate can be synthesized by the addition of zinc acetate dehydrate and glutaric acid. The structure is shown in Figure 29.5.

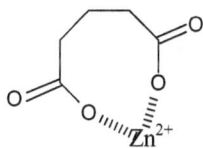

Figure 29.5: Structure of zinc glutarate.

Zinc glutarate has seen use as a catalyst by some polycarbonate producers. It can be prepared by addition of zinc salts to any of several glutarate sources, although the just-mentioned glutaric acid is the most common [15].

29.4 Catalyst fate

As with many catalysts discussed in earlier chapters, when the value of the metal in the catalyst is sufficient, efforts are made to recover the catalyst, even if it is in some way diminished or poisoned. The economic factors in recovering and reproducing the catalyst are the driving force. In cases where a metal is inexpensive, the catalyst maybe recovered if there is some cost or penalty in disposing of it, usually based on local or regional laws about pollution generation.

References

[1]   American Chemistry Council. Website. (Accessed 19 October 2020, as: www.americanchemistrycouncil.com/About).
[2]   The Polycarbonate/BPA Global Group. Website. (Accessed 20 October 2020, as: plastics. americanchemistry.com/Plastics/Product-Groups-and-Stats/PolycarbonateBPA-Global-Group/).
[3]   A&C Plastics, Inc. Website. (Accessed 20 October 2020, as: acplasticsinc.com/ informationcenter/r/types-of-polycarbonate-sheeting).
[4]   EC21, Global B2B Marketplace. Website. (Accessed 19 October 2020, as: ec21.com/ed-market/ US/nylon.html).
[5]   Plastics Europe. Website. (Accessed 19 October 2020 as: plasticseurope.org/en).
[6]   Chemistry Industry Association of Canada. CIAC. Website. (Accessed 19 October 2020, as: canadianchemistry.ca).
[7]   Chemicals and Petrochemicals Manufacturer's Association, India. CPMAI. Website. (Accessed 19 October 2020, as: www.cpmaindia.com/nfy_about.php).
[8]   Plastics Industry Manufacturers of Australia. PIMA. Website. (Accessed 20 October 2020, as: www.pima.asn.au).
[9]   Plastics SA. Website. (Accessed 20 October, as: www.plasticsinfo.co.za).
[10] Plastics New Zealand. Website. (Accessed 20 October 2020, as: www.plastics.org.nz).
[11] Japanese Patent. Fujita, K., Natsume, Y., and Ogasawara, T. Production of titanium tetrachloride. JPH0226828A, 1988.

[12] U.S. Patent. Ammons, V.G. Catalyst for making polycarbonate diols for use in polycarbonate urethanes. US4160853A, 1977.

[13] U.S. Patent. Okamoto, M., Sugiyama, J.-I., and Ueda, M. Catalyst for polycarbonate production and process for producing polycarbonate. US7390868B2.

[14] U.S. Patent. Meng, Y., Zhu, Q., Zhang, S., Li, X., and Du, L. Supported catalysts for the fixation of carbon dioxide into aliphatic polycarbonates and a process for preparing the same. US6844287B2, 2001.

[15] Ree, M., Hwang, Y., Kim, J.S., Kim, H., Kim, G., and Kim, H. New findings in the catalytic activity of zinc glutarate and its application in the chemical fixation of CO_2 into polycarbonates and their derivatives, 2006. Catalysis Today, 115, 1, 134–145.

Chapter 30
Polyester, polyethylene terephthalate (PETE)

30.1 Introduction

The broad class of materials is called polyesters, which take their name from the ester functionality repeated in the backbones in their polymer chains. In Chapter 26, we have already discussed nylon, the first polymer which is structurally akin to polyester, a polyamide. Thus, the entire history of polyester polymers has a definite start point that being the early polyamides first created when Dr. Carothers did his work in the 1930s. The history of the two types of polymers is intertwined, with Carothers doing early work on polymers made from diols and dicarboxylic acids but diverting his energies to polyamides, which gave rise to nylon. British scientists John Rex Whinfield, James Tennet Dickson, W.K. Birtwhistle, and C.G. Ritchie produced an early polyester at the outbreak of the Second World War, named Terylene [1–3]. The rights to this were purchased by DuPont, which produced an early polyester known as Dacron.

One of the largest industrial-scale polymerizations is the formation of the polyester polyethylene terephthalate (PETE) which is given the resin identification code 1 (RIC 1 or 01) – and which has been known as Terylene and Dacron. The repeat unit of PETE is shown in Figure 30.1. Annually, production is so large on a global scale that several trade organizations that advocate for the use of various plastics are in some way involved in monitoring and promoting its use, and often its recycling [4–8].

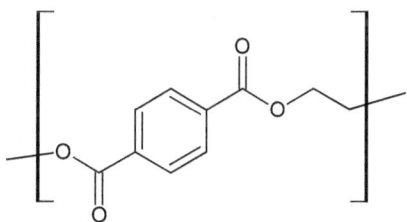

Figure 30.1: Polyethylene terephthalate repeat unit.

PETE is a thermoplastic polymer and finds its use in a wide variety of products and applications, including fabrics and bottles. In producing fibers, such a polymer is forced while hot through small holes in a metal plate. The resulting filaments then cool swiftly as their surface area is now high. It is widely recycled, and the RIC code can be seen on beverage bottles as well as many other food packaging items. Figure 30.2 gives one example. But there are many derivatives of the repeat structure shown in Figure 30.1, since different moieties can be placed where the phenyl ring is

https://doi.org/10.1515/9783110542868-030

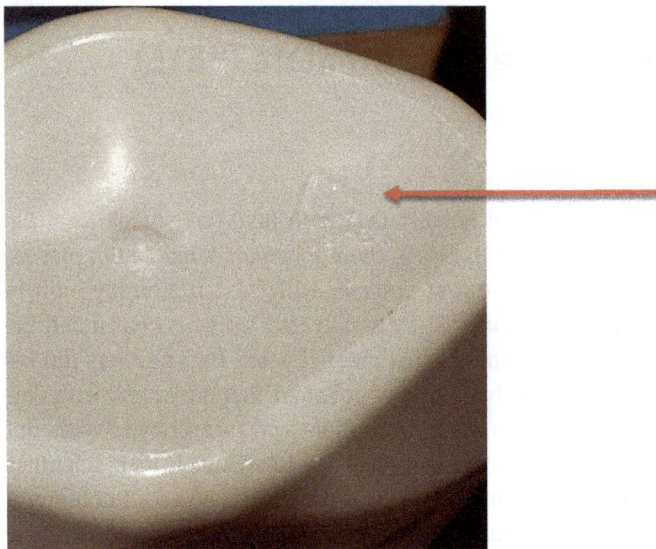

Figure 30.2: RIC 1 labeling on the base of a bottle.

positioned, both aromatic and aliphatic, and since different moieties can be placed where the ethyl group is located, again either aromatic or aliphatic.

Although PETE is the most common of the polyesters, this class of materials has become so large since its inception that some polyesters do not bear any RIC, or bear the RIC 7, meaning "other" plastics besides the six most commonly used. This diversity of structures and resulting polymers means that a discussion of catalysts also becomes broad. In this chapter and several of the following, we will focus on catalysts used in the production of the most common versions of the title polymers.

30.2 Reaction chemistry

All polyesters proceed via the same general synthetic route but can begin from different classes of starting materials. As already discussed in Chapter 26, the polymer forms along with some coproduct, often water, although HCl is also a possibility if the starting materials include chlorides. The reaction chemistry is much like that of the production of nylon but can also proceed through the opening of a cyclic structure which becomes the sole monomer. Figure 30.3 shows the basic condensation chemistry.

Note that when two starting materials are used, as mentioned, the reaction is a condensation, which is called so because of the water made as a by-product. Perhaps obviously, when starting materials are not diols and carboxylic acids, the by-product will not be water.

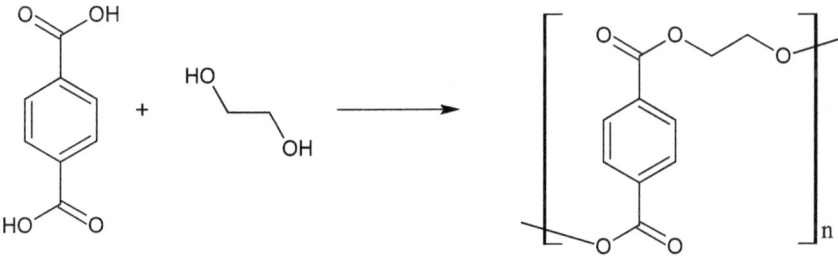

Figure 30.3: Polyethylene terephthalate production.

30.3 Catalyst production

Antimony trioxide has been used extensively as a catalyst for the production of polyester to increase molecular weight of the resulting polymer. To a lesser extent, titanium dioxide and germanium dioxide have used. This is because of their effects on the final polymer, enhancing its clarity, for example.

30.3.1 Antimony trioxide (Sb_2O_3, aka antimony (III) oxide)

This catalyst can be produced by the direct addition of the elemental metal to oxygen, as shown in Figure 30.4.

$$3O_{2(g)} + 4Sb \rightarrow 2\ Sb_2O_3$$

Figure 30.4: Antimony trioxide production.

This process is run at elevated temperature, and the product forms via sublimation. If Sb_2O_3 is formed from stibnite ore, it can be separated from arsenic impurities in the ore by a high-temperature separation – by boiling the crude ore. As well, Figure 30.5 illustrates how antimony trioxide can be isolated from stibnite ore.

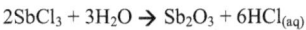

$$Sb_2S_3 + O_{2(g)} + CaCl_2 \rightarrow CaSO_4 + SbCl_3$$

$$2SbCl_3 + 3H_2O \rightarrow Sb_2O_3 + 6HCl_{(aq)}$$

Figure 30.5: Antimony trioxide production from stibnite.

Antimony materials are used largely as a flame retardant and as a pigment, with their use as a catalyst being somewhat smaller [9].

30.3.2 Antimony triacetate (Sb(C$_2$H$_3$O$_2$)$_3$)

This antimony compound is less widely utilized than its oxide counterpart, and routinely prepared from antimony trioxide, by reacting it with acetic acid, as shown in Figure 30.6, although acetic anhydride also has been proven to work in this capacity [10–12].

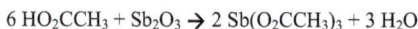

$$6 \, HO_2CCH_3 + Sb_2O_3 \rightarrow 2 \, Sb(O_2CCH_3)_3 + 3 \, H_2O$$

Figure 30.6: Production of antimony (III) acetate.

30.3.3 Titanium dioxide (TiO$_2$)

The production of titanium dioxide (aka, titanium (IV) oxide) has been discussed in Chapter 29 [13]. This material also finds extensive use as a pigment and as a whitener. It can be found in many foods, such as chewing gums. An example is shown in Figure 30.7.

Figure 30.7: Titanium oxide in a product ingredients list.

30.3.4 Germanium dioxide (GeO$_2$)

This compound also finds use in the formation of polyester [14] and can be produced via direct addition of germanium metal and oxygen. Germanium is largely imported to the United States and is found to be traces in ores when they are refined for other metals such as zinc ores [9].

30.4 Catalyst fate

Antimony compounds are routinely reclaimed if possible because of the value of the material. But several catalysts are incorporated into the plastic in the case of polyesters, especially if it is worked in the melt and the catalyst is in the mix.

Germanium is routinely recycled whenever it is economically feasible. The processes for doing so are well established, and the United States Geological Survey makes note that germanium is often recovered from "domestic industry-generated scrap" [9].

References

[1] Gaines, A. Wallace Carothers and the Story of DuPont Nylon: Unlocking the Secrets of Science.

[2] Hermes, M.E. Enough for One Lifetime: Wallace Carothers, Inventor of Nylon (History of Modern Chemical Sciences), ISBN: 978-0841233317, 1996.

[3] Whinfield, J.R. How Products Are Made. Website. (Accessed 27 October 2020, as: madehow. com/inventorbios/71/John-Rex-Whinfield.html).

[4] American Chemistry Council. Website. (Accessed 19 October 2020, as: www.americanchemistrycouncil.com/About).

[5] EC21, Global B2B Marketplace. Website. (Accessed 19 October 2020, as: ec21.com/ ed-market/US/nylon.html).

[6] Chemistry Industry Association of Canada. CIAC. Website. (Accessed 19 October 2020, as: canadianchemistry.ca).

[7] Plastics Europe. Website. (Accessed 19 October 2020 as: plasticseurope.org/en).

[8] Chemicals and Petrochemicals Manufacturer's Association, India. CPMAI. Website. (Accessed 19 October 2020, as: www.cpmaindia.com/nfy_about.php).

[9] USGS Mineral Commodity Summaries 2020. Downloadable as: pubs.er.usgs.gov/publication/ mcs2020.

[10] US Patent. Thomas, R.R. Preparation of antimony triacetate. US3415860A. 1966.

[11] US Patent. Otto Ernest Loeffler. Trivalent antimony catalyst. US3935170A, 1975.

[12] Chinese patent. Preparation method of antimony triacetate. CN1312102C, 2005.

[13] Schmidt, W., Thiele, U., Griebler, W.-D., Hirthe, B., and Hirschberg, E. Titanium-containing catalyst and process for the production of polyester. US5656716A, 1997.

[14] US Patent. Zoetbrood, G.J. Process of using germanium dioxide as a polyester condensation catalyst. US3,377,320, 1964.

Chapter 31
Polyethylene

31.1 Introduction

The simplest of the plastics produced by the activation of a double bond and production of a material from one starting molecule, polyethylene (PE), is produced to a total of over 100 million tons annually. The starting material, ethylene (C_2H_4), is an inexpensive material ultimately sourced from crude oil but can also be made by the dehydration of ethanol, as shown in Figure 31.1. The size of the production of PE is such that several national or regional trade organizations are devoted to its production and marketing, at least in part [1–10]. The production of polymers from it and from other small molecules that have a double bond at the terminal position is established enough that the processes are still sometimes called olefin polymerizations, or alpha-olefin polymerizations.

$$H_3C \diagdown_{OH} \longrightarrow H_2C{=}CH_2 + H_2O$$

Figure 31.1: Formation of ethylene from ethanol.

Since the increase to industrial levels of PE production, shortly after the Second World War, there has been an enormous amount of effort devoted to how it can be made with a wide variety of branches in the main chain – a factor that translates to how its density can be controlled. This research and development of new types and variations of PE means that there have been several improvements and expansions among the types of catalysts employed. To the consumer and general public, it also manifests itself in PE being assigned two resin identification codes (RICs) of the main six that are found on consumer end use products. Figures 31.2 and 31.3 show examples of plastic packaging which use RICs, namely RIC2 and RIC4.

    Note in Figure 31.2 that the item is high-density PE (HDPE). This refers to a type of PE that has minimal branching in its main carbon chains.

    Note in Figure 31.3 that the PE film packaging is low-density PE (LDPE), which means that there are significant amounts of branching in the main carbon chains. Both RICs are identifiers so that consumers or other end users may know they are recyclable plastics [11–16].

https://doi.org/10.1515/9783110542868-031

Figure 31.2: PE food container lid made from high-density polyethylene.

Figure 31.3: Low-density polyethylene food packaging.

31.2 Reaction chemistry

The synthesis of PE can be represented rather simply, as shown in Figure 31.4. It is difficult, however, to show the conditions that differentiate between the production of LDPE and HDPE. It involves the catalysts used – the focus of our discussion – as well as pressures and temperatures under which each type of material is produced.

Figure 31.4: Polyethylene production.

Since Figure 31.4 simply shows two repeat units of the starting monomer linked, it cannot show branching, or where it occurs in the production and growth of the polymer. As well, in no case can a reaction show the role of any of the catalysts used in producing PE.

31.3 Catalyst production

Phillips Petroleum developed one of the early means by which PE could be made on a large scale. The Phillips process produces HDPE and utilizes chromium trioxide, often on a silica support.

CrO_3 – sometimes called chromium trioxide, sometimes chromium (VI) oxide – is used for electroplating chromium onto surfaces and is used as a pigment. Thus, it is produced to function as a catalyst in what can be considered a secondary, but major use. Figure 31.5 shows the basic chemistry.

$$Na_2Cr_2O_7 + H_2SO_4 \rightarrow 2\ CrO_3 + H_2O + Na_2SO_4$$

Figure 31.5: Production of chromium trioxide.

The starting sodium dichromate in Figure 31.5 is itself a product of the refining of essentially all chromium ores, such as chromite. The United States imports roughly three-fourths of the chromium it uses annually from Kazakhstan, Russia, South Africa, and Mexico [17].

31.3.1 SiO$_2$ – silica

As mentioned in earlier chapters, silica is refined [18] from any number of sources, such as quartz. It is further purified for use as a catalyst, although silica is used widely in other materials such as different types of glass.

31.3.2 Ziegler–Natta catalysts

The expansion of PE production brought about a new class of catalysts that continues to expand even today. Karl Ziegler discovered that titanium-based catalysts promoted the polymerization of PE, and Giulio Natta expanded this to propylene polymerization. It was also found that magnesium chloride ($MgCl_2$) promoted and enhanced the polymerization of both ethylene and propylene.

31.3.3 Titanium tetrachloride on a support of magnesium chloride

31.3.3.1 TiCl$_4$

In Chapter 29, we discussed the production of titanium (IV) chloride, also known as titanium tetrachloride. The process often now begins with refined titanium dioxide

(TiO_2), and with the addition of both chlorine and elemental carbon (the latter to capture the bound oxygen) produces $TiCl_4$ and carbon monoxide.

31.3.3.2 MgCl$_2$

Magnesium chloride is used in a wide variety of applications beyond a catalyst support – such as magnesium metal production – and because of this, perhaps not surprisingly, can be produced in several different ways [17]. Liberation from an ammine complex is one, as shown in Figure 31.6.

$$[Mg(NH_3)_6]Cl_2 \rightarrow MgCl_2 + 6NH_3$$

Figure 31.6: Production of magnesium chloride.

The Dow process is another means of producing $MgCl_2$, one which utilizes hydrochloric acid to produce magnesium chloride, as shown in Figure 31.7.

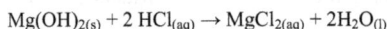

$$Mg(OH)_{2(s)} + 2\,HCl_{(aq)} \rightarrow MgCl_{2(aq)} + 2H_2O_{(l)}$$

Figure 31.7: Dow process for producing $MgCl_2$.

As mentioned, there are other means by which magnesium chloride can be produced, usually without a reduction to the elemental metal.

31.3.4 Metallocenes

The development of metallocene catalysts is a more recent addition to the large suite of catalysts that can polymerize PE (and other alpha-olefins). While there are now many metallocene catalysts, and while research into new ones continues, those that use titanium and zirconium, and to a lesser extent hafnium, have found wide use. Figure 31.8 gives a representative example.

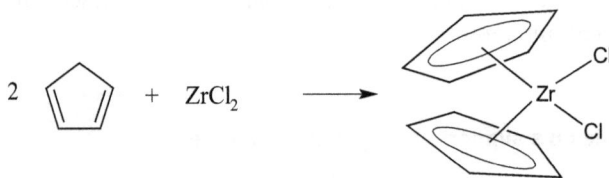

Figure 31.8: Example of a metallocene catalyst.

Such catalysts are routinely used with a cocatalyst, with methylaluminoxane being the most common [19–22].

Cyclopentadiene (Cp), the starting material for the synthesis of a metallocene, is actually recovered in relatively small quantities from the naphtha fraction of crude oil. Its direct synthesis from simpler starting materials requires several steps.

Pentamethylcyclopentadiene (abbreviated Cp*, and stated as, "C-P-star"), shown in Figure 31.9, is also used in metallocene catalysts. The presence of a methyl group on each carbon atom in the Cp ring prevents unwanted side reactions from occurring at the catalyst.

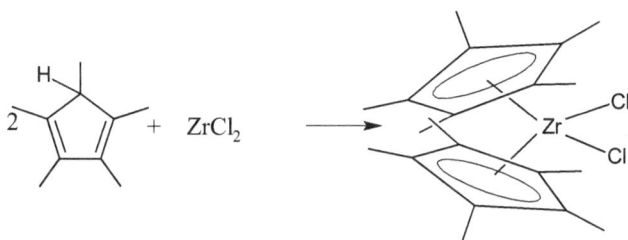

Figure 31.9: Example synthesis of a Cp* metallocene.

The original synthesis of Cp* is an established one [23] that requires several steps. Improvements have been made since the first reported synthesis of the molecule, although as with most organic materials, the ultimate source is some fraction of crude oil.

Additionally, ansa metallocenes have found their use as Ziegler–Natta catalysts. The term "ansa" means that a covalent bond exists between one ring of the metallocene and the other – a molecular bridge. Figure 31.10 shows an example, one in which a –CH₂– moiety connects the two rings. Many other connective pieces have been used as well, including some with two carbon atoms and some with silicon atoms in this bridge.

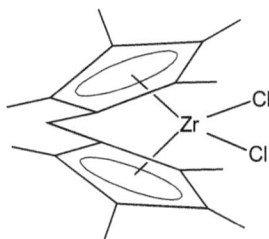

Figure 31.10: Example of an ansa metallocene.

Virtually all the metallocene-based catalysts are homogeneous, meaning they are in the same phase as that in which the polymerization occurs. This can make the recovery of the catalyst somewhat more difficult than if it were heterogeneous, especially if the catalyst were to be a solid while the reaction ran in solution or in the gas phase.

31.3.5 Methylaluminoxane cocatalyst (MAO)

This cocatalyst is used in the production of PE as well as other olefin polymerizations and is made from trimethylaluminum [22], as shown in Figure 31.11. In turn, the starting material trimethylaluminum is made from elemental metal and methylchloride, also shown in Figure 31.11.

$n\ H_2O\ +\ n\ Al(CH_3)_3 \rightarrow (Al(CH_3)O)_n\ +\ 2nCH_4$

from

$2Al\ +\ 6\ Na\ +\ 6CH_3Cl \rightarrow Al_2(CH_3)_6\ +\ 6\ NaCl$

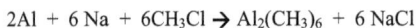

Figure 31.11: Synthesis of methylaluminoxane.

Note that trimethylaluminum is actually a dimer, despite this not being included in the chemical name. This is because aluminum salts tend to be hypovalent when compared to compounds and salts that obey the octet rule.

The combination of a metallocene catalyst and a MAO cocatalyst is often called a Kaminsky catalyst, or a Kaminsky catalyst system, after its discoverer, Professor Walter Kaminsky. The first reports of these catalysts date back to the early 1980s [24], and the research in this area is an ongoing effort in academic as well as corporate laboratories.

31.4 Catalyst fate

Heterogeneous catalysts are routinely easier to recover than homogeneous ones. This chapter discusses several of each broad type. But as with several of the other catalysts we have examined, the idea of recovery can be driven by the economic costs of the material that is the catalyst, the cost associated with proper disposal, as well as by the chemical and physical ease of reclamation. For example, Ziegler–Natta catalysts employing hafnium, while not particularly common, are rare enough that their recovery is usually advantageous.

References

[1] American Chemistry Council. Website. (Accessed 19 October 2020, as: www.americanchemis trycouncil.com/About).
[2] Plastics Industry Association. Website. (Accessed 1 November 2020, as: plasticsindustry.org).
[3] Manufacturers Association for Plastic Processors. Website. (Accessed 1 November 2020, as: www.mappinc.com).
[4] EC21, Global B2B Marketplace. Website. (Accessed 19 October 2020, as: ec21.com/ed-market/).
[5] Plastics Europe. Website. (Accessed 19 October 2020 as: plasticseurope.org/en).
[6] Chemistry Industry Association of Canada. CIAC. Website. (Accessed 19 October 2020, as: canadianchemistry.ca).
[7] Chemicals and Petrochemicals Manufacturer's Association, India. CPMAI. Website. (Accessed 19 October 2020, as: www.cpmaindia.com/).
[8] Plastics Industry Manufacturers of Australia. PIMA. Website. (Accessed 20 October 2020, as: www.pima.asn.au).
[9] Plastics SA (South Africa). Website. (Accessed 20 October, as: www.plasticsinfo.co.za).
[10] British Plastics Federation. Website. (Accessed 1 November 2020, as: bpf.co.uk).
[11] National Waste and Recycling Association. Website. (Accessed 1 November 2020, as: https:// wasterecycling.org).
[12] United states Environmental Protection Agency, "America Recycles. The U.S. Recycling System," Website. (Accessed 1 November 2020, as: https://www.epa.gov_us-recycling).
[13] Canadian Association of Recycling Industries. Website. (Accessed 1 November 2020, as: https://www.can-acir.org).
[14] The Recycling Association. Website. (Accessed 1 November 2020, as: recyclingassociation. com).
[15] Plastic Recyclers Europe. Website. (Accessed 1 November 2020, as: plasticsrecyclers.eu).
[16] Australian Council of Recycling. Website. (Accessed 1 November 2020, as: https://www.acor. org.au).
[17] U.S. G. S. Mineral Commodity Summaries, 2020. Downloadable as: https://pubs.er.usgs.gov/ mcs2020.
[18] U.S. Patent. Akira, S., Narita, Y., and Nagata, S. Process for Producing High Purity Silica, US4,973,462, 1989.
[19] Patent U.S. Fotis, P. Jr. Production of solid polyethylene by a catalyst consisting essentially of an alkali metal and an adsorbent alumina-containing material, US2,887,472, 1959.
[20] U.S. Patent. Lonfils, N., Bodart, P., and Debra, G. Catalysts for polyethylene production and use thereof, US 6,096,679, 2000.
[21] U.S. Patent. Ford, R.R., Ames, W.A., Dooley, K.A., Vanderbilt, J.J., and Wonders, A.G. Process for producing polyethylene, US6,191,239B1, 2001.
[22] Pullukat, T.J., and Hoff, R.E. Silica-based Ziegler-Natta catalysts: a patent review. Catalysis Reviews, 41, (1999), 389–428 and references therein.
[23] DeVries, L. Preparation of 1,2,3,4,5-pentamethyl-cyclopentadiene, 1,2,3,4,5,5-hexamethyl-cyclopentadiene, and 1,2,3,4,5-pentamethyl-cyclopentadienyylcarbinol. Journal of Organic Chemistry, 25, 10, (1960), 1838.
[24] Kaminsky, W. Highly active metallocene catalysts for olefin polymerization. Journal of the Chemical Society. Dalton Transactions, 9, (1998), 1413–1418, – and references therein.

Chapter 32
Polypropylene

32.1 Introduction

The chemistry of polypropylene (PP) is very much like that of polyethylene, discussed in Chapter 31, uses some of the same catalysts in certain conditions, and yet does have some important differences. The addition of a methyl group, pendant to each monomeric repeat unit in the polymer, may not seem like a large difference structurally. But this does produce three different types of PP, based on the arrangement and order, or disorder, of the pendant methyl groups. This is termed "tacticity," and the three forms are titled isotactic, syndiotactic, and atactic. The Lewis structures of the repeat units of the first two are shown in Figures 32.1 and 32.2. It is difficult to show the lack of tacticity – the atactic version – of PP by showing simply the repeat units of the Lewis structure. The term "atactic" means that there is no repeatability to the methyl side chains of this type of PP. Isotactic PP – sometimes abbreviated iPP – is the form overwhelmingly used in commercial end user applications.

As with polyethylene, PP is produced on such a large scale that there are several national or regional trade organizations that promote its uses and applications [1–11].

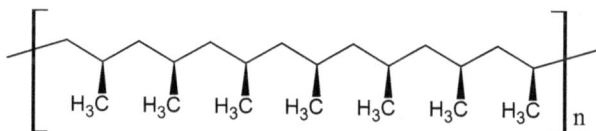

Figure 32.1: Isotactic polypropylene.

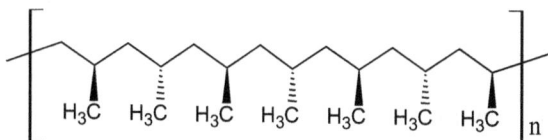

Figure 32.2: Syndiotactic polypropylene.

As with several other polymers that are produced on an industrial scale, PP has its own resin identification code (RIC). The code 5, or 5-PP, indicates PP, like the other RICs listed on consumer products so that it is known that the material can be recycled. Figure 32.3 shows an example from a prescription medicines container. Recycling of used plastics from some consumer end use item is large enough that

https://doi.org/10.1515/9783110542868-032

Figure 32.3: Example of RIC 5.

there are many regional and national organizations that promote and encourage recycling [12–19], much like there are organizations that promote the material's use.

32.2 Reaction chemistry

The reaction for the production of PP can be shown very simply, starting with propylene (C_3H_6), as in Figure 32.4.

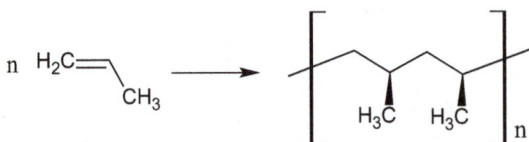

Figure 32.4: Production of polypropylene.

But again, this does not illustrate how the tacticity of the product influences its properties, and its larger form. iPP forms helices, which account for many of its desirable properties, such as hardness and relatively high melting point.

32.3 Catalyst production

Like the production of polyethylene, soluble Ziegler–Natta catalysts are used extensively in this polymerization. Indeed, much of the early work on the polymerization of propylene was influenced by, and some would say modeled after, the polymerization of polyethylene. Thus, many of the catalysts are the same.

TiCl$_4$

Al(C$_2$H$_5$)$_3$ – to increase activity of TiCl$_4$

MgCl$_2$ – as a support for TiCl$_4$

The production of these three catalysts or cocatalysts has been discussed in the last chapter, when showing how they are made for polyethylene production.

32.3.1 Diethylaluminum halide, (CH$_3$CH$_2$)$_2$AlX$_3$

This compound actually does not increase the activity of production of PP more than all other catalysts and cocatalysts. But it does provide an increase in the specific product, biasing it toward iPP. It is one of several in which the anion can be a halide or ethoxide [20, 21].

$$2\ Al\ +\ 3\ C_2H_5Cl \rightarrow (C_2H_5)_3AlCl_3$$

Figure 32.5: Production of an ethylaluminum halide.

Figure 32.5 shows one of the halides that are produced in large amounts. But it should be noted that while the fluoride does not give the greatest increase in activity, it does provide the greatest specificity in producing iPP.

32.3.2 Metallocenes

In Chapter 31, we showed the structure of a metallocene and of an ansa metallocene. In syndiotactic PP, the form used in smaller amounts than iPP, it has been found that ansa metallocenes incorporating one fluorenyl portion are effective in directing the polymerization, when combined with organoaluminum compounds such as methylaluminoxane (also discussed in Chapter 31). Figure 32.6 shows the basic structure.

Zirconium is not the only metal that can be used in such catalysts. Metals such as hafnium have been proven to be useful in this polymerization [22–24].

The synthesis of such metallocenes is a matter of forming the ansa-bridged organic portion first, then inserting the metal in the center of the "sandwich" or "bent sandwich" molecule.

$$X = -CH_2-, \text{ or } -Si(CH_3)_2-, \text{ or } -Si(C_6H_5)_2-, \text{ or } -CH_2CH_2-$$

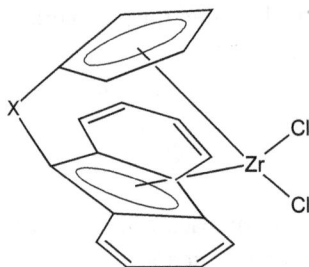

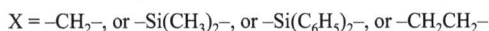

Figure 32.6: Ansa metallocene with fluorenyl ring.

32.4 Catalyst fate

As mentioned in Chapter 31, heterogeneous catalysts are easier to recover than homogeneous ones because they are in a different phase than the material being polymerized. Yet as is the case of many other catalysts, the idea of recycling a catalyst is routinely one that involves the cost of the recovery as one factor in an equation, and the inherent value of the catalyst – usually the metal – as the other factor.

Interestingly, catalysts used in the production of PP can be reclaimed by deashing the portion of the reaction that is not the product [25].

References

[1]   American Chemistry Council. Website. (Accessed 19 October 2020, as: www.americanchemis trycouncil.com/About).
[2]   Plastics Industry Association. Website. (Accessed 1 November 2020, as: plasticsindustry.org).
[3]   Manufacturers Association for Plastic Processors. Website. (Accessed 1 November 2020, as: www.mappinc.com).
[4]   EC21, Global B2B Marketplace. Website. (Accessed 19 October 2020, as: ec21.com/ed-market/).
[5]   Plastics Europe. Website. (Accessed 19 October 2020 as: plasticseurope.org/en).
[6]   Chemistry Industry Association of Canada. CIAC. Website. (Accessed 19 October 2020, as: canadianchemistry.ca).
[7]   International Association of plastics Distribution. Website. (Accessed 3 November 2020, as: iapd.org).
[8]   Industrial Quick Search Manufacturer Directory (IQS Directory). Polypropylene Manufacturers and Suppliers. Website. (Accessed 3 November 2020, as: iqsdirectory.com/polypropylene).
[9]   Chemicals and Petrochemicals Manufacturer's Association, India. CPMAI. Website. (Accessed 19 October 2020, as: www.cpmaindia.com/).
[10]   Plastics Industry Manufacturers of Australia. PIMA. Website. (Accessed 20 October 2020, as: www.pima.asn.au).

[11]   Plastics SA (South Africa). Website. (Accessed 20 October, as: www.plasticsinfo.co.za).
[12]   British Plastics Federation. Website. (Accessed 1 November 2020, as: bpf.co.uk).
[13]   National Waste and Recycling Association. Website. (Accessed 1 November 2020, as: https://
       wasterecycling.org).
[14]   United states Environmental Protection Agency, "America Recycles. The U.S. Recycling
       System," Website. (Accessed 1 November 2020, as: https://www.epa.gov_us-recycling).
[15]   Canadian Association of Recycling Industries. Website. (Accessed 1 November 2020, as:
       https://www.can-acir.org).
[16]   The Recycling Association. Website. (Accessed 1 November 2020, as: recyclingassociation.
       com).
[17]   Plastic Recyclers Europe. Website. (Accessed 1 November 2020, as: plasticsrecyclers.eu).
[18]   Australian Council of Recycling. Website. (Accessed 1 November 2020, as: https://www.acor.
       org.au).
[19]   The Association of Plastics Recyclers. Website. (Accessed 3 November 2020, as:
       plasticsrecycling.org).
[20]   U.S. Patent. Ewen, J.A. Catalyst for producing hemiisotactic polypropylene. US5,036,034,
       1991.
[21]   U.S. Patent. McDaniel, M.P., Shveima, J.S., Benham, E.A., Geerts, R.L., and Smith,
       J.L. Polymerization catalyst systems and processing using alkyl lithium compounds as a
       cocatalyst. US6,828,268B1, 2004.
[22]   Ewen, J.A., Jones, R.L., and Razavi, A. Syndiospecific propylene polymerization with group 4
       metallocenes. Journal of the American Chemical Society, 110, (1988), 6255–6256.
[23]   U.S. Patent. Klendworth, D.D., Reinking, M.K., Kist, E.W., and Meyer, K.E. Propylene
       polymerization process with enhanced catalyst activity. US6,630,544B1, 2003.
[24]   U.S. Patent. Ommundsen, E., Follestad, A., Harkonen, M., Poikela, M., Jaaskelainen, P., and
       Alastalo, K. Process for producing propylene based polymer compositions, US6,770,714B2,
       2004.
[25]   U.S. Patent. Di Petro, J., and Millington, W. Process for de-ashing polymers. US3,560,471,
       1971.

Chapter 33
Polystyrene

33.1 Introduction

Polystyrene (often abbreviated PS) is another plastic or polymer made on an enormous scale each year (tens of millions of tons), and dependent upon the opening of one double bond. Figure 33.1 shows the starting monomer, styrene, to emphasize that it is what is called the exocyclic double bond – the double bond not in the phenyl ring – that is opened in the polymerization from styrene to PS. Styrene itself is produced from benzene via an alkylation to produce ethylbenzene, then a subsequent dehydrogenation. As might be expected, several of the world's largest petroleum producers all manufacture styrene. Also as might be expected, there exist numerous trade associations and organizations that promoted the use of PS, as well as other plastics that are produced on a similar scale [1–13].

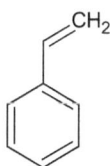

Figure 33.1: Lewis structure of styrene.

Unlike monomers such as ethylene and propylene, styrene is made from two hydrocarbons, which in turn come from crude oil, benzene, and ethylene, as shown in Figure 33.2. Also, the intermediate ethylbenzene must be dehydrogenated to obtain the styrene monomer [14].

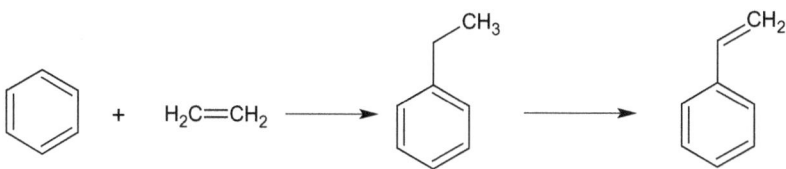

Figure 33.2: Production of styrene.

Both the production of ethylbenzene, and the subsequent production of styrene, each require a catalyst, as well as elevated temperature and pressure. Zeolites are the favored catalysts for the production of ethylbenzene (specifically ZSM-5), and iron (III) oxide has been well established for the dehydrogenation of ethylbenzene to styrene.

https://doi.org/10.1515/9783110542868-033

Much like with polypropylene (PP) or polyvinyl chloride, a discussion of tacticities of PS involved three possibilities. Isotactic PS, as shown in Figure 33.3, has all phenyl groups on the same side of the carbon atom main polymer chain, syndiotactic PS has the phenyl groups alternating from side to side, and atactic PS shows no order in the placement of side chain phenyl groups. Atactic PS is by far the most common form of PS and is used in almost all consumer end use items such as packaging, although patents have been filed for the production of syndiotactic PS [15, 16].

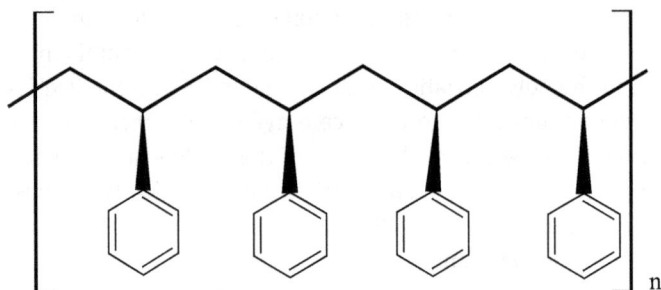

Figure 33.3: Isotactic polystyrene.

In discussing the tacticity possibilities of PS, it is logical to mention Styrofoam®, the material produced by Dow Chemical. The name is trademarked by Dow Chemical, but the term has been widely adopted by the general public for any sort of PS material used as a lightweight, low-density insulator, in packaging, and in cups, to name just two of a multitude of uses. Its extremely low density is the result of a great deal of air being present in the user end product. This is also largely responsible for its insulating ability.

For most individuals, the resin identification code 6 – RIC6 PS, as shown in Figures 33.4 and 33.5 – is associated not only with Styrofoam but with any user end product that is made of this lightweight material. The material is recyclable, and many organizations exist to promote the recycling of plastics [17–25]. But the extreme low density of PS or Styrofoam often makes it economically unfavorable to transport to some recycling center.

The RIC6 PS code is international, although in Japan the inner symbol is in a square, and says "pura māku," meaning "recyclable plastic."

33.2 Reaction chemistry

The reaction for the manufacture of PS can be shown in a simple form, starting with styrene ($C_6H_6CHCH_2$), as in Figure 33.6.

But once again, a simplified reaction does not illustrate how the tacticity or atacticity of the product polymer influences its macroscopic properties. Atactic PS is the first that was patented and brought to the consumer market in any fashion and remains the major type of PS that is produced [26].

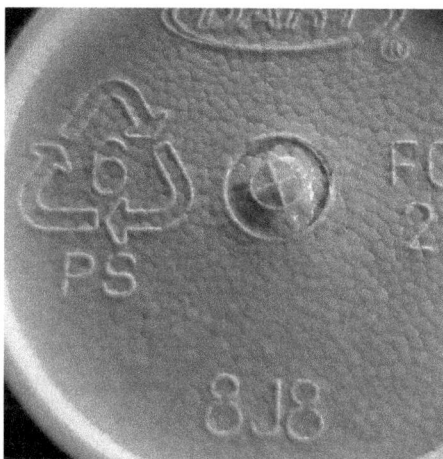

Figure 33.4: Polystyrene RIC.

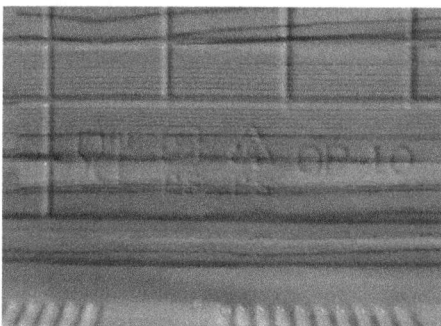

Figure 33.5: Polystyrene recycle symbol in western lettering and in Japanese.

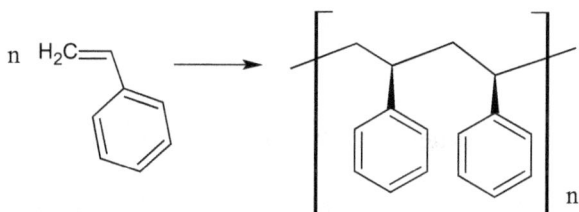

Figure 33.6: Production of polystyrene.

33.3 Catalyst production

33.3.1 For styrene

33.3.1.1 Zeolite ZSM-5

There is a large class of materials called zeolites, often aluminosilicates, but ZSM-5, first produced by Mobil Oil, is now common enough that it is produced by chemical companies for sale to research laboratories, as well as to industrial concerns which use it as a catalyst [27]. Figure 33.7 shows the simplified reaction chemistry by which it can be made, although more than one method exists.

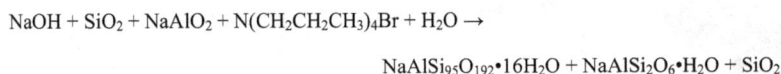

$$NaOH + SiO_2 + NaAlO_2 + N(CH_2CH_2CH_3)_4Br + H_2O \rightarrow$$

$$NaAlSi_{95}O_{192} \cdot 16H_2O + NaAlSi_2O_6 \cdot H_2O + SiO_2$$

Figure 33.7: Production of ZSM-5.

The reaction, while aqueous, does require elevated temperature and pressure. Also, it requires an inert reaction vessel or liner. Polytetrafluoroethylene has been used successfully in this regard.

33.3.1.2 Iron (III) oxide, aka ferric oxide (Fe₂O₃)

We have mentioned in previous chapters that iron (III) oxide is used largely in the iron and steel industry but can be made in a pure form in a variety of ways. Figure 33.8 shows the reaction using sodium hydroxide as a reactant.

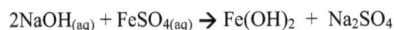

$$2NaOH_{(aq)} + FeSO_{4(aq)} \rightarrow Fe(OH)_2 + Na_2SO_4$$

followed by

$$Fe(OH)_2 + O_{2(g)} \rightarrow Fe_2O_3 + H_2O$$

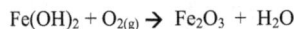

Figure 33.8: Production of iron (III) oxide.

33.3.2 For polystyrene

Other choices can be used for the production of PS from styrene, especially since the atactic form is that which is used the most. Early patented processes [26] use benzoyl peroxide as a radical initiator, shown in Figure 33.9, in a catalytic fashion, and also discuss the use of Fuller's earth and activated clay as supports.

More recently, metallocene catalysts, much like those used in the production of polyethylene and PP, have been employed in the production of PS. Methylaluminoxane

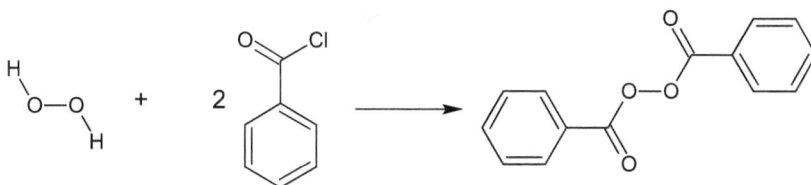

Figure 33.9: Benzoyl peroxide.

cocatalysts, discussed in previous chapters on olefin polymerizations, are also employed but efforts have been made to find means by which they can be eliminated [28, 29].

33.4 Catalyst fate

An early problem in the production of PS was being able to extract any metal catalyst from the product. Benzoyl peroxide decomposes and is not recovered. But newer catalysts, such as the metallocenes, must be extracted from the product to ensure the highest purity of PS. This is one reason that research into catalysts for PS production continues, as it remains desirable to utilize and perfect a system that has no catalyst or cocatalyst to recover or extract [28].

References

[1]   American Chemistry Council. Website. (Accessed 19 October 2020, as: www.americanchemis trycouncil.com/About).
[2]   Plastics Industry Association. Website. (Accessed 1 November 2020, as: plasticsindustry.org).
[3]   Manufacturers Association for Plastic Processors. Website. (Accessed 1 November 2020, as: www.mappinc.com).
[4]   EC21, Global B2B Marketplace. Website. (Accessed 19 October 2020, as: ec21.com/ed-market/).
[5]   Plastics Europe. Website. (Accessed 19 October 2020 as: plasticseurope.org/en).
[6]   Chemistry Industry Association of Canada. CIAC. Website. (Accessed 19 October 2020, as: canadianchemistry.ca).
[7]   International Association of Plastics Distribution. Website. (Accessed 3 November 2020, as: iapd.org).
[8]   Industrial Quick Search Manufacturer Directory (IQS Directory). Polypropylene Manufacturers and Suppliers. Website. (Accessed 4 November 2020, as: iqsdirectory.com/polystyrene).
[9]   The Japan Plastics Industry Federation. Website. (Accessed 4 November 2020, as: jpif.gr.jp/english).
[10] Chemicals and Petrochemicals Manufacturer's Association, India. CPMAI. Website. (Accessed 19 October 2020, as: www.cpmaindia.com/).

[11]   Plastics Industry Manufacturers of Australia. PIMA. Website. (Accessed 20 October 2020, as: www.pima.asn.au).

[12]   Plastics SA (South Africa). Website. (Accessed 20 October, as: www.plasticsinfo.co.za).

[13]   British Plastics Federation. Website. (Accessed 1 November 2020, as: bpf.co.uk).

[14]   The Essential Chemistry – online, Poly(phenylethene) (Polystyrene). Website. (Accessed 4 November 2020, as: essentialchemicalindustry.org/polymers/polyphenylethene.html).

[15]   U.S. Patent. Okada, A. Syndiotactic polystyrene composition. US5352727A, 1994.

[16]   U.S. Patent. Ooki, Y. Thermoplastic elastomer resin composition and connector. US2012/0094539A1, 2012.

[17]   National Waste and Recycling Association. Website. (Accessed 1 November 2020, as: https://wasterecycling.org).

[18]   United states Environmental Protection Agency, "America Recycles. The U.S. Recycling System," Website. (Accessed 1 November 2020, as: https://www.epa.gov_us-recycling).

[19]   Canadian Association of Recycling Industries. Website. (Accessed 1 November 2020, as: https://www.can-acir.org).

[20]   The Recycling Association. Website. (Accessed 1 November 2020, as: recyclingassociation.com).

[21]   Plastic Recyclers Europe. Website. (Accessed 1 November 2020, as: plasticsrecyclers.eu).

[22]   Australian Council of Recycling. Website. (Accessed 1 November 2020, as: https://www.acor.org.au).

[23]   The Association of Plastics Recyclers. Website. (Accessed 3 November 2020, as: plasticsrecycling.org).

[24]   Plastics NZ: New Zealand's Industry Association. Website. (Accessed 4 November 2020, as: plastics.org.nz/environment/recycling-disposal).

[25]   South African Plastics Recycling Organisation SAPRO. Website. (Accessed 4 November 2020, as: www.plasticrecyclingsa.co.za).

[26]   U.S. Patent. Wakeford, L.E., and Helmsley, D. Manufacture of polystyrene. US2,556,488, 1951.

[27]   U.S. Patent. Argauer, R.J., and Landolt, G.R. Crystalline zeolite ZSM-5 and method of preparing the same. US3702886, 1975.

[28]   European Patent. Ewen, J.A., and Elder, M.J. Addition of aluminum alkyl for improved metallocene catalyst, EP0,426,638A2, 1990.

[29]   U.S. Patent. Turner, H.W. Polymers from ionic metallocene catalyst compositions. US6,232,420B1, 2001.

Chapter 34
Polytetrafluoroethylene

34.1 Introduction

The initial production of polytetrafluoroethylene (PTFE) by accident has become one of the classic stories in which serendipity is recognized by an inquiring mind. Dr. Roy Plunkett, working for DuPont, discovered in 1938 that tetrafluoroethylene stored in a metal container self-polymerized, leaving a waxy, slick substance as the product – PTFE. The product has since been called Teflon®, and the term has joined a few other brand names and trade names to become a household word (words such as Kleenex®, Xerox copy, or Styrofoam®). Figure 34.1 shows the simplified chemical reaction for the polymerization [1, 2].

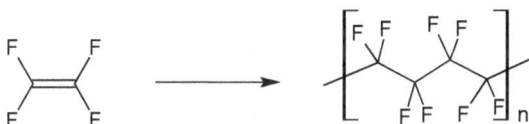

Figure 34.1: Production of PTFE.

Note that the reaction chemistry is much like that of polyethylene production, but with fluorine atoms in place of each of the hydrogen atoms on the initial monomer. In short, the reaction chemistry is again the reactivity of the olefin double bond.

34.2 Reaction chemistry

Since the initial production of PTFE, significant effort has been made to optimize conditions and find the best performing catalysts. The original patent [1] tried several catalysts, including zinc chloride, $ZnCl_2$, as well as silver nitrate, $AgNO_3$ in several parts, as well as $AgNO_3$ and methyl alcohol, as well as benzoyl peroxide. The original claims were that yields were highest with silver (I) nitrate, or with silver (I) nitrate and methyl alcohol [1].

More recently, PTFE has been produced in containers that are stainless steel, and has used initiators such as ammonium persulfate ($NH_4S_2O_8$), sometimes still called ammonium peroxydisulfate.

All the different catalysts that have been used to produce PTFE still enable the same reaction, that shown in Figure 34.1.

https://doi.org/10.1515/9783110542868-034

34.3 Catalyst production

34.3.1 Zinc (II) chloride, ZnCl$_2$

Hydrochloric acid can be used to produce zinc chloride by the following reaction:

$$Zn_{(s)} + 2HCl_{(aq)} \rightarrow ZnCl_{2(aq)} + H_{2(g)}$$

The oxidation of zinc by acid is straightforward, and industrially the only other step in the synthesis is the removal of water from the product. The product is a white powder.

34.3.2 Silver (I) nitrate, AgNO$_3$

Silver shot or silver foil can be used to prepare silver nitrate. The other reactant that must be present is nitric acid.

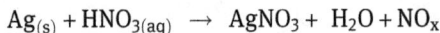

$$Ag_{(s)} + HNO_{3(aq)} \rightarrow AgNO_3 + H_2O + NO_x$$

Silver nitrate remains one of the least expensive salts of silver, and is one of the most soluble. Thus, it finds many uses besides that of a catalyst. It exists as a white solid.

34.3.3 Methanol, CH$_3$OH

The production of methanol from methane first requires the production of $CO_{(g)}$ from methane, what is often called syn gas. Next, the CO, which is often in a product mixture with CO_2 undergoes what is called the water–gas shift reaction, which produces methanol. A simplified reaction is

$$7H_2 + CO + CO_2 \rightarrow 2CH_3OH + 2H_2 + H_2O$$

Methanol is used in a large number of reactions, and its use here in a catalytic role is because it was specifically mentioned in the first PTFE patent. Its synthesis was described in more detail in Chapter 24.

34.3.4 Ammonium persulfate, NH$_4$S$_2$O$_8$

The production of ammonium persulfate is an electrolytic process, depending on the combination of ammonium sulfate in sulfuric acid:

$$(NH_4)_2SO_4 + H_2SO_4 \rightarrow NH_4S_2O_8$$

All the starting materials for this production are themselves inexpensive enough that the cost of the catalyst is small when compared to the cost of other materials, and of the energy input, ultimately required to produce PTFE.

References

[1] U.S. Patent No. 2,230,654, Tetrafluoroethylene Polymers, February 4th, 1941.
[2] Puts, G.J., Crouse, P., and Ameduri, B.M. Polytetrafluoroethylene: synthesis and characterization of the original extreme polymer. Chemical Reviews, 119, 3, (2019), 1763–1805.

Chapter 35
Polyvinyl chloride (PVC)

35.1 Introduction

Polyvinyl chloride (PVC) is another of the major plastics produced in the world, third only to polyethylene and polypropylene in amounts produced. Tens of millions of tons are produced annually, with the material going to a wide variety of consumer end products, the most familiar being pipes and tubing. Since it is a commodity plastic, the trade organizations that advocate for and promote other plastics, such as polyethylene, polypropylene, polystyrene, and others also advocate for the use of PVC [1–14]. The general public tends to think of PVC piping as its major use, and indeed, because of how well it resists exposure to chemicals, it finds extensive use in aboveground and underground piping. But there are many other consumer end uses for PVC.

The starting material for PVC is the vinyl chloride monomer (VCM), discussed in a later chapter. Of all the alpha olefin monomers used for bulk plastic production, this is the only one we have seen that requires some atom that does not come from crude oil – namely, the chlorine. Figure 35.1 shows the Lewis structure of vinyl chloride.

Figure 35.1: Vinyl chloride.

Curiously, the single chlorine atom in the monomer accounts for nearly 57% of the mass of the monomer, and thus of the mass of the polymer. The basic math is as follows:

$$1\ Cl = 35.453\ amu$$
$$2\ C = 24.022\ amu$$
$$3\ H = 3.023\ amu$$
$$Total = 62.498$$
$$Cl/total = 35.453/62.498 = 0.5673\ or\ 56.73\%.$$

One aspect this polymer shares with propylene and styrene is that the pendant group, in this case the chlorine atom, influences the macroscopic properties of the resulting polymer. The possibility of an isotactic PVC, as well as a syndiotactic PVC, and an atactic PVC all exist. In practical terms, almost all PVC that is produced on an industrial scale is atactic PVC and has plasticizers added to it, often esters. There have been some reports of syndiotactic PVC [15, 16], but the large majority of it is still produced in an atactic form, in part because this form was the first to be brought up to large scale [17].

https://doi.org/10.1515/9783110542868-035

Like the other plastics produced on an enormous scale, PVC can be recycled. The national and regional recycling associations are the same as those which focus on plastics such as polyethylene, polypropylene, and polystyrene [18–25]. PVC carries number 3 as a resin identification code, and on user end, items can be abbreviated PVC or V. It can be chemically or mechanically recycled [26].

35.2 Reaction chemistry

Vinyl chloride, or the VCM, is discussed later in this volume. Virtually all vinyl chloride is used to produced PVC, as shown in Figure 35.2.

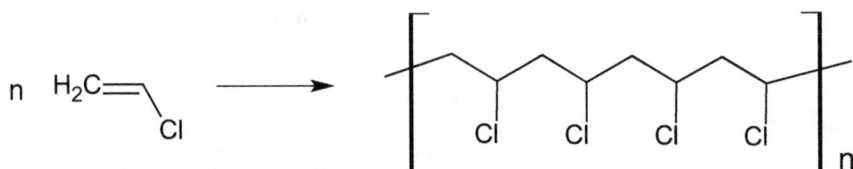

Figure 35.2: Polymerization of vinyl chloride.

We have not shown any three-dimensional tacticity at the chlorine–carbon bonds in Figure 35.2, because, as mentioned, most PVC is atactic.

35.3 Catalyst production

The polymerization of vinyl chloride proceeds via radical polymerization, with some organic peroxide often used as an initiator, which basically functions in a catalyst-like role. Since PVC is produced on such a large scale, and through so many different companies, it is difficult to state that a particular peroxide is the primary one for the process. For example, Figure 35.3 shows two peroxides that have seen considerable use in producing PVC.

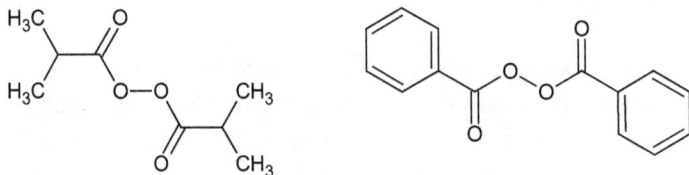

Figure 35.3: Diisobutyryl peroxide and benzoyl peroxide.

Although benzoyl peroxide tends to be a well-known peroxide among the general public, because of its uses in such applications as acne medication, peroxides such as the diisobutyryl peroxide are made in large enough quantities that they have trade names. For example, Nouryon markets this peroxide as Trigonox 187-WD40 [27].

Benzoyl peroxide is produced through the addition of hydrogen peroxide and benzoyl chloride, as shown in Figure 35.4.

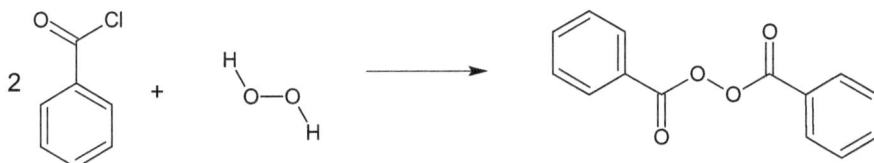

Figure 35.4: Benzoyl peroxide production.

In turn, benzoyl chloride can be produced by the reaction of water with benzotrichloride, which is itself produced by the chlorination of toluene. Since toluene is one of the molecules distilled from crude oil or produced from benzene and *para*-xylene, the ultimate source for this peroxide is crude oil and its refinement.

35.4 Catalyst fate

Unlike the catalysts that have been discussed which include some metal, often a precious metal, the organic peroxide initiators used in PVC production are degraded in the polymerization process, and thus are not recovered.

References

[1]   American Chemistry Council. Website. (Accessed 19 October 2020, as: www.americanchemis trycouncil.com/About).
[2]   Plastics Industry Association. Website. (Accessed 1 November 2020, as: plasticsindustry.org).
[3]   Manufacturers Association for Plastic Processors. Website. (Accessed 1 November 2020, as: www.mappinc.com).
[4]   EC21, Global B2B Marketplace. Website. (Accessed 19 October 2020, as: ec21.com/ed-market/).
[5]   Plastics Europe. Website. (Accessed 19 October 2020 as: plasticseurope.org/en).
[6]   Chemistry Industry Association of Canada. CIAC. Website. (Accessed 19 October 2020, as: canadianchemistry.ca).
[7]   International Association of Plastics Distribution. Website. (Accessed 3 November 2020, as: iapd.org).

[8] Industrial Quick Search Manufacturer Directory (IQS Directory). PVC Manufacturers and Suppliers. Website. (Accessed 6 November 2020, as: https://www.iqsdirectory.com/pvc/pvc–2/).

[9] The Japan Plastics Industry Federation. Website. (Accessed 4 November 2020, as: jpif.gr.jp/english).

[10] Chemicals and Petrochemicals Manufacturer's Association, India. CPMAI. Website. (Accessed 19 October 2020, as: www.cpmaindia.com/).

[11] Plastics Industry Manufacturers of Australia. PIMA. Website. (Accessed 20 October 2020, as: www.pima.asn.au).

[12] Plastics SA (South Africa). Website. (Accessed 20 October, as: www.plasticsinfo.co.za).

[13] British Plastics Federation. Website. (Accessed 1 November 2020, as: bpf.co.uk).

[14] → The Essential Chemistry – online, Poly(phenylethene) (Polystyrene). Website. (Accessed 4 November 2020, as: essentialchemicalindustry.org/polymers/polyphenylethene.html).

[15] U.S. Patent. Threads, fibers, fabrics, tapes and films based on post-chlorinated polyvinyl chloride, and method for production thereof. R. Buening, K.-H. Diessel, H.E. Konermann. US3,462,402, 1969.

[16] U.S. Patent. Flame resistant, thermally stable polyvinyl chloride – polyester blends. W. Darchuk. US4,004,405, 1977.

[17] U.S. Patent. Polymerization catalyst for vinyl chloride. L.F. Marous, C.D. McCleary. US3,022,282, 1962.

[18] National Waste and Recycling Association. Website. (Accessed 1 November 2020, as: https://wasterecycling.org).

[19] United states Environmental Protection Agency, "America Recycles. The U.S. Recycling System," Website. (Accessed 1 November 2020, as: https://www.epa.gov_us-recycling).

[20] Canadian Association of Recycling Industries. Website. (Accessed 1 November 2020, as: https://www.can-acir.org).

[21] The Recycling Association. Website. (Accessed 1 November 2020, as: recyclingassociation.com).

[22] Plastic Recyclers Europe. Website. (Accessed 1 November 2020, as: plasticsrecyclers.eu).

[23] Australian Council of Recycling. Website. (Accessed 1 November 2020, as: https://www.acor.org.au).

[24] Plastics NZ: New Zealand's Industry Association. Website. (Accessed 4 November 2020, as: plastics.org.nz/environment/recycling-disposal).

[25] South African Plastics Recycling Organisation SAPRO. Website. (Accessed 4 November 2020, as: www.plasticrecyclingsa.co.za).

[26] EcoMENA Echoing Sustainability in MENA. Website. (Accessed 8 November 2020, as: ecomena.org/recycling-pvc/).

[27] Nouryon Polymer Chemistry. Website. (Accessed 8 November 2020, as: polymerchemistry.nouryon.com/product/trigonox-187-wd40-diisobutyryl-peroxide-cas-3437-84-1/).

Chapter 36
Propylene

36.1 Introduction

This small olefin is the lowest molecular weight hydrocarbon – again, alkene or olefin – that can produce polymers of three different tacticities. The reason for this is the possible arrangements of the methyl groups at the double bond after polymerization. Propylene is produced as a coproduct with ethylene during the steam cracking process of a wide variety of hydrocarbon feed stocks. As well, it can be produced by the dehydrogenation of propane, something the major oil companies are heavily invested in [1–10]. But it can also be produced from bio-based feedstocks. Its structure is shown in Figure 36.1.

Figure 36.1: Lewis structure of propylene.

Like the other small molecules that function as monomers for the polymers produced on the largest scale, propylene and the different forms or polypropylene are materials that are the focus of several regional or national trade associations [11–22]. In general, trade associations tend to be centered around all plastics, and not simply one plastic.

36.2 Reaction chemistry

Propylene is routinely produced from the dissociation of ethylene and 2-butene, as shown in Figure 36.2. The well-established method for this production is the Phillips Triolefin Conversion, and yields over 90% propylene. While this process has been used for decades, research continues on the development of catalysts related to propylene production [23–26].

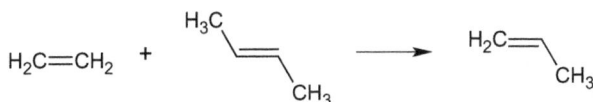

Figure 36.2: Means of production of propylene.

https://doi.org/10.1515/9783110542868-036

Figure 36.3: RIC5, polypropylene.

Essentially, all propylene is used to produce polypropylene, a plastic made on such a large scale that it has a dedicated resin identification code of 5. Figure 36.3 shows an example of this on a recyclable household product.

This example is from a beverage container lid, but there are many other possible examples as well. Polypropylene is widely recycled.

36.3 Catalyst production

While a wide variety of catalysts have been tried for the production of propylene, and numerous supports, such as silica, alumina, and zeolites, have also been used to stabilize such catalysts, those materials listed below have a long, established history in this role.

36.3.1 Rhenium oxide (aka, dirhenium heptoxide)

Re_2O_7, known more formally as rhenium (VII) oxide, can be formed through direct combination of the elements at elevated temperature. Figure 36.4 shows its formation from other oxides or sulfides at elevated temperature, generally 500–700 °C.

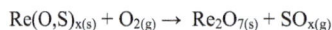

$$Re(O,S)_{x(s)} + O_{2(g)} \rightarrow Re_2O_{7(s)} + SO_{x(g)}$$

Figure 36.4: Dirhenium heptoxide formation.

The reaction is not shown balanced, since formation is dependent upon the composition of the original rhenium-bearing source.

Rhenium is a rare enough commodity that it is tracked by the United States Geological Survey (USGS) and is counted "in kilograms of rhenium content [27]," whereas most of the commodities we have discussed in this book are tracked in tons, and in some cases millions of tons of annual production. Further, the USGS indicates that "Most rhenium occurs with molybdenum in porphyry copper deposits [27]." Thus, all rhenium is recovered as a by-product in the refining of other metals.

36.3.2 Molybdenum oxide (MoO₃)

The production of molybdenum oxide can be from any molybdenum ore, but is most commonly produced from molybdenum (IV) sulfide, called molybdenite when extracted as an ore. Figure 36.5 shows the basic reaction chemistry.

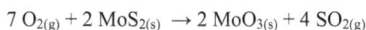

$$7\,O_{2(g)} + 2\,MoS_{2(s)} \rightarrow 2\,MoO_{3(s)} + 4\,SO_{2(g)}$$

Figure 36.5: Molybdenum oxide formation.

Molybdenum is a more common metal than rhenium, and is the primary element mined in some operations throughout the world. It can also be recovered as a secondary product from copper porphyry ores. Much of the molybdenum oxide that is formed is later reduced to the metal, then added as a component of steel in its manufacturing. Thus, its use as a catalyst is not its major use; but its use in steel makes it an important commodity tracked by national governments [27–30].

36.4 Catalyst fate

Any type of catalyst that contains a transition metal of the rarity of rhenium will be recovered for reuse whenever possible. Likewise, metals such as molybdenum are recycled whenever possible, simply because the ease of recycling is far greater than the costs involved in extracting such metals from newly mined ores [27].

References

[1]   China Petroleum & Chemical Corporation (Sinopec). Website. (Accessed 18 November 2020, as: www.sinopec.com).
[2]   PetroChina Company, Ltd. Website. (Accessed 18 November 2020, as: www.petrochina.com. cn).
[3]   Saudi Arabian Oil Company (Saudi Aramco). Website. (Accessed 18 November 2020, as: www. aramco.com).
[4]   Royal Dutch Shell PLC. Website. (Accessed 18 November 2020, as: www.shell.com).

[5]   BP PLC. Website. (Accessed 18 November 2020, as: bp.com).
[6]   Exxon Mobil Corporation. Website. (Accessed 18 November 2020, as: corporate.exxonmobil. com).
[7]   Total SE, France. Website. (Accessed 18 November 2020, as: total.com).
[8]   Chevron Corporation. Website. (Accessed 18 November 2020, as: www.chevron.com).
[9]   Marathon Petroleum Corporation. Website. (Accessed 18 November 2020, as: https://www. marathonpetroleum.com).
[10]   PJSC Lukoil (LUKOY). Website. (Accessed 18 November 2020, as: Lukoil.com).
[11]   American Chemistry Council. Website. (Accessed 19 October 2020, as: www.americanchemis trycouncil.com/About).
[12]   Plastics Industry Association. Website. (Accessed 1 November 2020, as: plasticsindustry.org).
[13]   Manufacturers Association for Plastic Processors. Website. (Accessed 1 November 2020, as: www.mappinc.com).
[14]   EC21, Global B2B Marketplace. Website. (Accessed 19 October 2020, as: ec21.com/ed-market/).
[15]   Plastics Europe. Website. (Accessed 19 October 2020 as: plasticseurope.org/en).
[16]   Chemistry Industry Association of Canada. CIAC. Website. (Accessed 19 October 2020, as: canadianchemistry.ca).
[17]   International Association of Plastics Distribution. Website. (Accessed 3 November 2020, as: iapd.org).
[18]   The Japan Plastics Industry Federation. Website. (Accessed 4 November 2020, as: jpif.gr.jp/ english).
[19]   Chemicals and Petrochemicals Manufacturer's Association, India. CPMAI. Website. (Accessed 19 October 2020, as: www.cpmaindia.com/).
[20]   Plastics Industry Manufacturers of Australia. PIMA. Website. (Accessed 20 October 2020, as: www.pima.asn.au).
[21]   Plastics SA (South Africa). Website. (Accessed 20 October, as: www.plasticsinfo.co.za).
[22]   British Plastics Federation. Website. (Accessed 1 November 2020, as: bpf.co.uk).
[23]   European Patent Application. Leyshon, D.W., Sofranko, J.A., and Jones, C.A. Production of propylene from higher hydrocarbons. EU90310719.1, 1990.
[24]   International Patent. Sawyer, G.A. Propylene production. WO 2010/077263 A2, 2010.
[25]   U.S. Patent. Hood, A.D. Jr., and Bridges, R.S. Staged propylene production process, US9309168B2, 2016.
[26]   Del Campo, P., et al. Propene production by butene cracking: descriptors for zeolite catalysts. ACS Catalysis, 10, 20, (2020), 11878–11891.
[27]   U.S.G.S. Mineral Commodity Summaries 2020. Downloadable as: pubs.er.usgs.gov/ publication/mcs2020.
[28]   Minerals UK: Centre for sustainable mineral development. Directory of Mines and Quarries. Website. (Accessed 19 November 2020, as: www2.bgs.ac.uk/mineralsUK/mines/dmq.html).
[29]   Government of Canada. Minerals and Mining Publications. Website. (Accessed 19 November 2020, as: nrcan.gc.ca/maps-tools-publications/publications/minerals-mining-publications/18733).
[30]   Study on the EU's list of Critical Raw Materials (2020). Downloadable as: CRM_2020_Non-Critical%20Factsheets.pdf.

Chapter 37
Propylene oxide and propylene glycol

37.1 Introduction

In this chapter, we treat two commodity chemicals as a single subject, because they are so closely related in the means by which they are produced, in their ultimate starting material, and in their end uses. The Lewis structures for propylene oxide and propylene glycol are shown in Figure 37.1. Note that propylene glycol also goes by the name 1,2-propane diol. It has one chiral center, but in all industrial-scale applications, it is used as a racemic mixture.

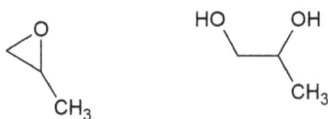

Figure 37.1: Lewis structures of propylene oxide and propylene glycol.

Propylene glycol is produced primarily for use in making several different commercially valuable polymers, but does have other uses. For example, it has the number E-1520 when used as a food additive. In this capacity, it is an emulsifier or thickener, and can be found in a wide variety of products, such as salad dressings, cake mixes, and other packaged foods. It is often used to help the food maintain its flavor for longer periods of time than if it were not added. Figure 37.2 shows one example.

Propylene oxide is routinely produced from propene (propylene), and has other uses besides the production of propylene glycol, often the production of polyurethanes. Yet, it is often used for the production of propylene glycol. Since both are the precursors for commodity plastics, the same organizations that advocate for the use of such plastics have an interest in the production of these two materials. Thus, there are several national and regional trade organizations that take interest in these materials [1–13].

37.2 Reaction chemistry

It would be logical to presume that the production of propylene oxide simply mirrors that of ethylene oxide, also treated in this volume, in Chapter 18. This is not the case, however. Propylene oxide has traditionally been made by a very different synthetic pathway. Because this method coproduces a by-product, significant effort has

https://doi.org/10.1515/9783110542868-037

Figure 37.2: Propylene glycol in a food product's ingredients list.

been made and continues to be made for an economically viable alternative, one that does not produce any chlorinated by-product [14].

The production of propylene oxide via two different routes is shown in Figures 37.3 and 37.4. Notice that in the first method, what is called propylene chlorohydrin is produced, the chlorine of which must then be removed in a reaction termed dehydrochlorination. The base used to provide the hydroxide is often calcium hydroxide or sodium hydroxide. The choice of base is usually one of economics, meaning a particular base is the most available and least expensive. This means that calcium chloride or sodium chloride becomes a by-product. While this may seem complex for an industrial-scale process, it has become a very established one.

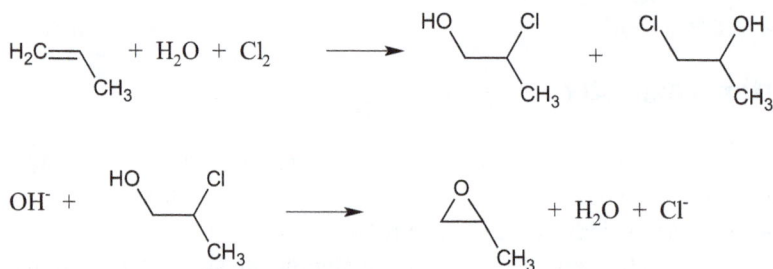

Figure 37.3: Production of propylene oxide.

The second method, shown in Figure 37.4, avoids the use of a chlorinated intermediate. This however does require some organic peroxide, such as *t*-butyl hydroperoxide [15]. Hydrogen peroxide can be used, in what is termed the hydrogen peroxide to propylene oxide process (the HPPO process). This process, developed by Evonik and ThyssenKrupp International, does require a titanium catalyst in the form of a zeolite. Because it does not need any chlorine, emphasis has been placed on its environmental friendliness [16, 17].

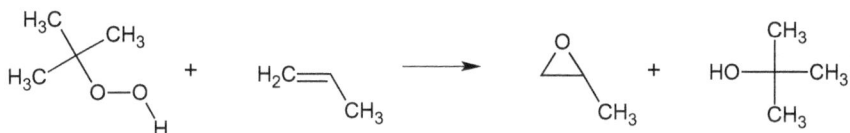

Figure 37.4: Production of propylene oxide via peroxide.

The production of propylene glycol from propylene oxide can be performed at approximately 150 °C using a broad array of catalysts. The catalyst can be sulfuric acid, can be an alkali, or can be an ion exchange resin. Figure 37.5 shows the basic reaction chemistry.

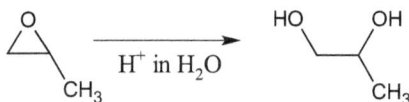

Figure 37.5: Production of propylene glycol from propylene oxide.

37.3 Catalyst production

The production of new catalysts aimed at better conversion and more efficient production of these two products continues, including even those based on precious metals as a component of the catalyst [14]. But those shown here are well established, and have found use on an industrial scale.

37.3.1 TS-1, titanium silicate

This catalyst, for the production of propylene oxide, is a zeolite, with tetrahedral titanium dispersed into silica, and is ultimately produced from SiO_2 and TiO_2 [17, 18]. The Kroll process has traditionally been the method by which titanium is obtained as a source material. The simplified reaction chemistry for this is shown in Figure 37.6.

$$TiO_2 + C + Cl_2 \rightarrow TiCl_4 + CO_2$$

$$TiCl_4 \rightarrow 2Mg \rightarrow 2\,MgCl_2 + Ti$$

Figure 37.6: Kroll process.

While the Kroll process is shown here starting with titanium (IV) oxide – rutile ore – it can also utilize $FeTiO_3$ – ilmenite ores.

37.3.2 Sulfuric acid, catalytically

We treat sulfuric acid in Chapter 40, and mention that it is the commodity chemical most produced in the world. A trade organization exists to promote the use of sulfuric acid, and to discuss improvements in production [19]. Ultimately, sulfuric acid is produced from sulfur, oxygen, and water, with one step requiring V_2O_5 as a catalyst.

37.3.3 Alkali, for propylene glycol production

The production of alkalis such as NaOH or KOH is done via well-established routes. The chlor-alkali process for the production of sodium hydroxide is shown in Figure 37.7.

$$2NaCl_{(aq)} + 2H_2O_{(l)} \rightarrow 2NaOH_{(aq)} + Cl_{2(g)} + H_{2(g)}$$

Figure 37.7: Chlor-alkali process.

37.4 Catalyst fate

Since the hydrochlorination method of propylene oxide preparation produces an enormous amount of a chloride such as calcium chloride or sodium chloride, finding some use for these by-products has always been of greater concern than treatment of catalysts. Should silver-based catalysts displace this widely on an industrial scale, the recovery of spent silver catalysts will become economically important.

Sulfuric acid and sodium hydroxide are inexpensive enough that their recovery is a matter of economic concern, in terms of weighing the cost of recovery against any possible cost associated with disposal.

References

[1] American Chemistry Council. Website. (Accessed 19 October 2020, as: www.americanchemis-trycouncil.com/About).

[2] Plastics Industry Association. Website. (Accessed 1 November 2020, as: plasticsindustry.org).

[3] Manufacturers Association for Plastic Processors. Website. (Accessed 1 November 2020, as: www.mappinc.com).

[4] EC21, Global B2B Marketplace. Website. (Accessed 19 October 2020, as: ec21.com/ed-market/).

[5] Plastics Europe. Website. (Accessed 19 October 2020 as: plasticseurope.org/en).

[6] Chemistry Industry Association of Canada. CIAC. Website. (Accessed 19 October 2020, as: canadianchemistry.ca).

[7] International Association of Plastics Distribution. Website. (Accessed 3 November 2020, as: iapd.org).

[8] Industrial Quick Search Manufacturer Directory (IQS Directory). Polypropylene Manufacturers and Suppliers. Website. (Accessed 4 November 2020, as: iqsdirectory.com/polystyrene).

[9] The Japan Plastics Industry Federation. Website. (Accessed 4 November 2020, as: jpif.gr.jp/english).

[10] Chemicals and Petrochemicals Manufacturer's Association, India. CPMAI. Website. (Accessed 19 October 2020, as: www.cpmaindia.com/).

[11] Plastics Industry Manufacturers of Australia. PIMA. Website. (Accessed 20 October 2020, as: www.pima.asn.au).

[12] Plastics SA (South Africa). Website. (Accessed 20 October, as: www.plasticsinfo.co.za).

[13] British Plastics Federation. Website. (Accessed 1 November 2020, as: bpf.co.uk).

[14] Ghosh, S., Acharyya, S.S., Tiwar, R., Sarkar, B., Singha, R.K., Pendem, C., Sasaki, T., and Bai, R. Selective oxidation of propylene to propylene oxide over silver-supported tungsten oxide nanostructure with molecular oxygen. ACS Catalysis, 4, 7, (2014), 2169–2174.

[15] Kollar., J. US Patent. Catalytic epoxidation of an olefinically unsaturated compound using an organic hydroperoxide as an epoxidizing agent. US3,350,422, 1967.

[16] HPPO Technology – An eco-friendly and cost effective direct synthesis of propylene oxide. YouTube. (Accessed 24 November 2020).

[17] Tsuji, J., Yamamoto, J., Ishino, M., and Oku, N. Development of a New Propylene Oxide Process, R&D Report, "Sumitomo Kagaku," 2006-1.

[18] European Patent. Seo, T., and Abekawa, H. Method for producing propylene oxide. EP 2,014,654A1, 2007.

[19] Sulfuric Acid Today. Website. (Accessed 23 November 2020, as: h2so4today.com).

Chapter 38
Rubber

38.1 Introduction

In the large universe of polymers, let us begin by defining rubber – a polymer whose utility is manifested by being flexible and stretchable, and returns (more or less) to its original shape after deformation. Technically, their use is above their glass transition temperature, at which molecular mobility changes from highly mobile (above) to reduced mobility below – often a transition from "soft" to "hard or rigid." Often these polymers are cross-linked by controlled reaction to render their usefulness in an article more consistent or stable.

Any chapter on rubber is remiss without a brief mention of natural rubber, the substance which is produced from the latex or harvested sap of the Amazonian rubber tree *Hevea brasiliensis* [1], with this species preferred as it grows well in warm, humid climates with proper cultivation and tending. This has been available to humanity since antiquity. Much of the world's natural rubber is harvested in Thailand and Indonesia and its primary chemical composition is equivalent to that of *cis*-polyisoprene [1] containing a few residual impurities from its biogenesis (phosphates, enzymes, etc.) as shown in Figure 38.1. Its natural sourcing has been both boon and bane – market supply depending on climate and yield, harvesting resources, and global political stability. It was this latter consideration that drove researchers to explore the creation of synthetic rubber. For example, during the Second World War, Indonesian supply to both the Allied Forces and Axis Powers became constrained.

Prior to these forced supply crises, Fritz Hofmann invented synthetic polyisoprene at the Bayer Labs in Germany in 1909 [2], and Russian Sergey Lebedev had created polybutadiene. Lebedev in 1930, German Hermann Staudinger in 1931, and DuPont researcher Wallace Carothers independently invented poly-2-chlorobutadiene, later sold successfully as the heat- and oil-resistant Neoprene® elastomer. In about 1935, German researchers developed the family of elastomers later known as Bunas. Understandably,

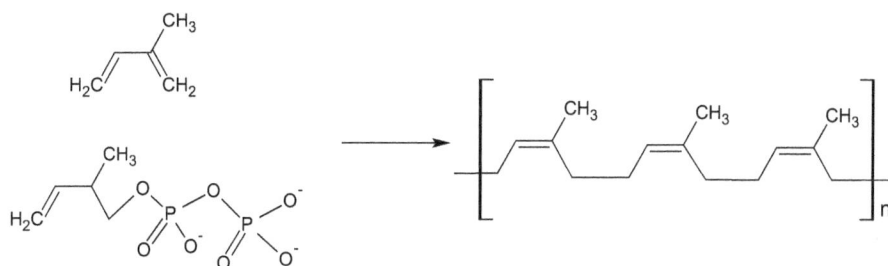

Figure 38.1: Natural polyisoprene structure.

https://doi.org/10.1515/9783110542868-038

the focus of these efforts was directed to free nations from dependency upon harvested natural rubber from Asia.

Still widely used in tires, footwear, and many articles of trade, natural rubber, the structure shown in Figure 38.1, remains plagued by the global supply and demand fluctuations and pricing.

38.2 Synthetic rubbers and use in tires

Approximately 300 million new vehicle tires are consumed annually in North America [3] containing about 1 billion kilograms of synthetic rubber. A tire is among the most highly engineered but undervalued composites, comprising four to six chemically distinct rubber or elastomeric polymers, steel, and other fiber belting, and considerable amounts of reinforcing, often carbon black or functionalized silica, as shown in Figure 38.2 (figure courtesy of G4sxe). Although there are numerous other applications of synthetic rubber (hoses, sealants, adhesives, etc.), we will focus on those requiring catalysis in the dominant tire market. An automobile tire of ca. 12 kg contains about 4 kg of synthetic rubber, and about 2 kg each of natural rubber and carbon black. The tread is often made of *cis*-polybutadiene or polyisoprene (wear and skid resistant), sidewall and carcass of emulsion-polymerized styrene butadiene rubber, inner liner of air-impermeable butyl or halobutyl rubber and the outer sidewall of ethylene–propylene–diene (EPDM) copolymer to impart ozone resistance [3]. A number of these are appropriate for inclusion in this book as their synthesis requires catalytic processes.

We will discuss each of these rubber classes in turn [3]. The predominant use of catalysis in rubber is in the polymerization process, usually with transition metal or rare earth Ziegler–Natta chemistry. The catalysts are never recovered or recycled as the quantities employed are so low (the catalyst has been extracted, not for recovery but to impart color stability of products in subsequent processes).

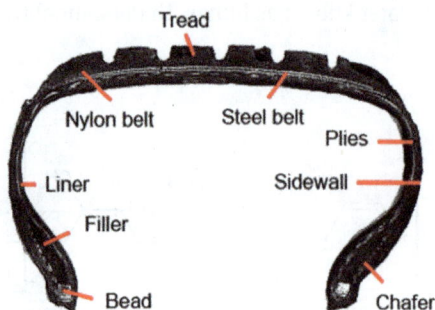

Figure 38.2: Cross section of passenger tire construction [4, 5].

38.2.1 Polyisoprene

To obtain a rubber with the excellent properties akin to natural rubber, researchers turned their focus to synthesis of polydienes, in particular *cis*-polyisoprene, as a chemical clone of natural rubber. Polyisoprene is prepared commercially using several methods, each of which produces of mixtures of structures in the polymer chain. These are usually called microstructures, shown in Figure 38.3, originating from the various ways in which the 1,2- and 1,4 polymerizations may occur.

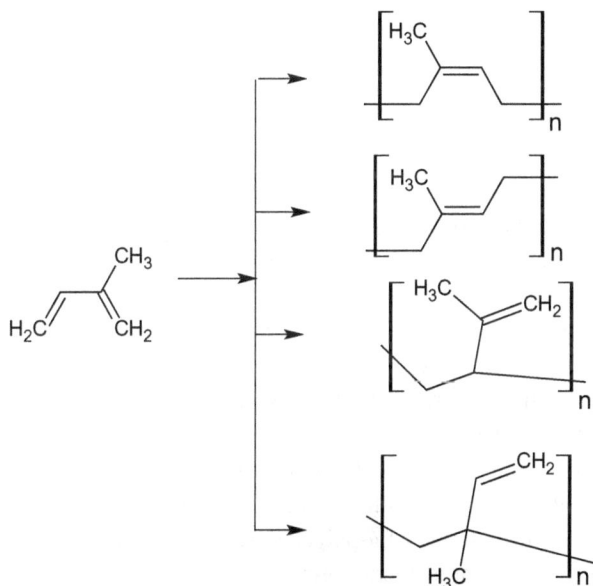

Figure 38.3: Synthetic polyisoprene microstructures.

In the case of anionic *n*-butyllithium-initiated polymerization in hydrocarbon solution, over 90% of the groups in the polymer chain are *cis*-1,4, about 3% are *trans*-1,4 and the remainder 3,4 units. This has profound effects on the final application properties but in general produces a rubber very close to its natural analogue.

In the cases of Ziegler–Natta catalysts, the microstructure in rubber is quite different and can be divided into three broad categories:
1. Coordinative chain polymerization, this catalysis uses a titanium tetrachloride mix with triisobutylaluminum ($TiCl_4/Al(i\text{-}C_4H_9)_3$), resulting in a purer *cis*-1,4-polyisoprene. This is much like natural rubber [6].
2. In the case of vanadium (III) chloride and triisobutylaluminum as the catalysts ($VCl_3/Al(i\text{-}C_4H_9)_3$), the predominant result is *trans*-dominant polyisoprene [6]. This material is a stiff, thermoformable plastic used for example in golf ball covers.

3. When the catalyst used is MO_2Cl_2 supported with a phosphorus ligand and a co-catalyst of $Al(O(C_6H_5)CH_3)(i\text{-}C_4H_9)_2$, both 1,2 and 3,4 are the dominant resulting polyisoprene [7].

As the catalysts are of high activity and turnover, there are generally no efforts made to recover and recycle them.

With the limited availability of the monomer isoprene from refinery and crude oil cracking, the polyisoprene rubber is not the synthetic rubber of choice in tire construction.

38.2.2 Polybutadiene

The annual production of near the turn of the millennium, polybutadiene production had topped two million tons. This is second only to styrene–butadiene rubber (SBR) as far as industrial production by volume [8, 9].

Similar to polyisoprene, polybutadiene can be produced commercially using several systems. In addition, it has the same microstructures available as those in polyisoprene: *cis*-1,4 and *trans*-1,4 and 1,2- (also called vinyl), as seen in Figure 38.4. The *cis*-1,4 shows best performance in tires, with a better mix of characteristics: tread wear, rolling resistance, and traction. A great advantage beyond performance is available from the monomer which is usually isolated in large quantities from C4 fraction from refinery operations. Butadiene is able to polymerize via three different means. These are titled: *cis*, *trans*, and *vinyl*. When butadiene units connect end to end, the result is the *cis* and *trans* forms. This is termed 1,4-polymerization. Different physical properties are associated with the different isomeric forms. "High *cis*"-polybutadiene, for example, has good elasticity which makes it preferred in tire construction. "High *trans*" is a crystalline plastic possessing limited useful applications. The third isomer, the vinyl content, is usually not more than a few percent. Beyond this, both the molecular weights and the amount of branching of polybutadienes do differ.

Besides their use in tire construction, large volumes of polybutadiene are used in producing high-impact polystyrene (HIPS) in which the rubber is dissolved in styrene where it is polymerized and grafted via a free radical initiator. The resulting product contains rubber microdomains imparting crack resistance and improved impact resistance. HIPS is used in many articles, food trays, toys, and kitchen appliances.

During polymerization, the *trans*- double bonds create generally straight polymer main chains. This rigidity results in microcrystalline regions in the polymer material. Any *cis*- double bonds produce polymer chain bends. These bends prevent the polymer chains from aligning, and thus do not form crystalline regions. In turn, this produces significant regions of amorphous or flexible polymer. Enhanced elastomeric (rubber-like) qualities are seen in polymers with significant amounts of the *cis*- double bond arrangement in their make-up.

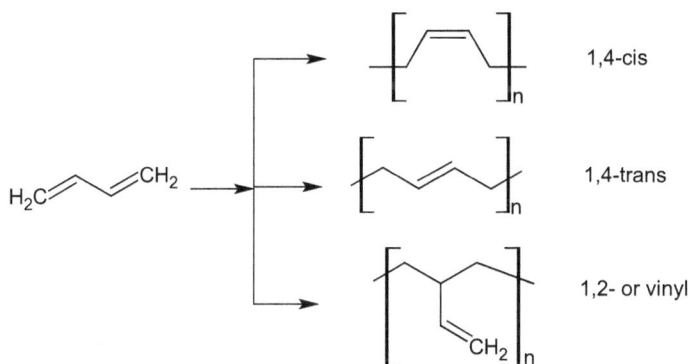

Figure 38.4: Polybutadiene microstructures.

The catalyst used in their production significantly affects the utility via micro-structure of polybutadiene product [8–10].

Table 38.1: Polybutadiene microstructure by catalyst/initiator [8–10].

Catalyst elemental basis	1-4-*Cis* content (mol%)	1-4-*Trans* content (mol%)	1-2-Vinyl content (mol%)
Neodymium	98	1	1
Cobalt	96	2	2
Nickel	96	3	1
Titanium	93	3	4
Lithium	10–30	20–60	10–70

As shown in Table 38.1, metals solubilized as 2-ethylhexanoates (octoates) or naph-thenates are usually in divalent form. Lithium in the form of *n*-butyl lithium acts as an initiator consumed in the living polymerization rather than a catalyst in the true sense.

38.2.2.1 High cis polybutadiene

Generally, high *cis*-polybutadiene contains more than 92% *cis* repeating units, and less than 4% vinyl units. Commonly, Ziegler–Natta catalysts incorporating neodymium or cobalt, as well as other metals, are used in its production. Physical properties of the polymer do vary somewhat depending on the choice of metal [9]. Typically, the polymerization is carried out in hydrocarbon solutions (such as butene, cyclohexane, benzene) using both the solubilized metal (anions such as octoate) and reducing co-catalysts (organoaluminum compounds like diethyl aluminum chloride (DEAC)).

The utilization of a cobalt-based catalyst solubilized with organics results in branched molecules. This tends to be a material of low viscosity, but also of relatively low mechanical strength. A catalyst based on neodymium produces the most linear overall structure, with correspondingly higher *cis* percentages, up to 98%. This also results in increased mechanical strength. Titanium- and nickel-based catalysts are less used.

38.2.2.2 Low cis polybutadiene

When the catalyst is an alkyllithium – such as butyl-lithium – the resulting polybutadiene product is termed, "low *cis*." Such material typically contains about 60% *trans*- roughly 30% *cis*- and the remaining percentage vinyl- internal structures.

Low *cis*- material is often combined with different tire polymers, even though it has a liquid–glass transition temperature that is considered high.

38.2.2.3 High vinyl polybutadiene

Utilizing an alkyllithium initiator, it was found, as early as 1980 by the company Zeon, that high-vinyl polybutadiene – having greater than 70% vinyl units – could be combined successfully with high *cis* in tires [9]. This material still possesses a high liquid–glass transition temperature.

38.2.2.4 High trans polybutadiene

Greater than 90% *trans*- units in polybutadiene can be manufactured using neodymium, lanthanum, or nickel-based catalysts much like those which produce high *cis*-polybutadiene. The resulting polymeric material can be characterized as a plastic crystal, and thus is not considered an elastomer. Such materials have melting points near 80 °C. It was formerly used for the outer layer of golf balls. Current uses for such are exclusively industrial [9].

38.2.2.5 Metallocene-catalyzed polybutadiene

Japanese researchers appear to be at the leading edge in the use of metallocene-based catalysts for butadiene polymerizations [11]. The perceived benefits could be first an enhanced degree of control concerning molecular mass distribution, and second the control of cis/trans/vinyl percentages. Metallocene catalyst-based polybutadiene is currently not manufactured on a commercial basis.

38.2.3 Ethylene–propylene–diene rubbers

Ethylene–propylene copolymers as well as terpolymers – generally abbreviated EPM and EPDM respectively – are also practical, functional polymers [12]. The former is used in applications such as adhesives, sealants, roofing membranes, and water isolation and containment membranes [13]. The saturated copolymer EPM is easily processable but can only be cross-linked by virtue of inclusion of peroxides or high-energy irradiation, without the presence of any pendant unsaturation point. Thus, the inclusion of an unsaturated diene up to 4% molar composition allows co-vulcanization into other compounded rubbers (see Tire construction 38.3). Typical dienes include these byproducts from refinery operations (Figure 38.5).

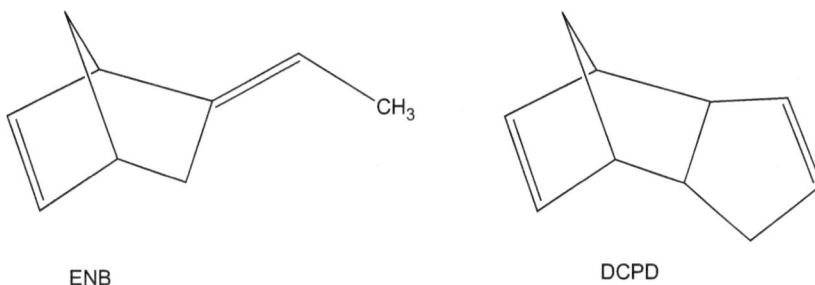

ENB   DCPD

Figure 38.5: 1,5-Ethylidene-2-norbornadiene (ENB) or dicyclopentadiene (DCPD).

These elastomers have good chemical resistance, especially to ozone, making them useful in tire sidewall construction. About 500 kt of EPDM are produced annually in North America [12, 14].

Akin to other polyolefins, their commercial production employs a transition metal catalyst like vanadium coupled with a suitable organometallic reducing agent like triethyl aluminum or DEAC [13].

38.2.4 Others

There are many other synthetic rubbers, none requiring true catalytic processes for their production. A quick overview, solely for the purpose of completeness will be provided here.

38.2.4.1 Styrene–butadiene

Styrene butadiene is a commodity copolymer used in many durable goods, serving largely as a carcass elastomer in tire construction. Much of the rubber is produced in an emulsion polymerization process using a redox initiation system and not a bona fide catalyst [14]. These rubbers have a very random microstructure. SBR can also be produced using an anionic process with *n*-butyl lithium as the in initiator, yielding a "living polymer" and block-polymer architecture with a rigid, plastic-like segment or block of predominantly styrene, a flexible rubbery segment of butadiene and finally the last block being styrene again. These blocks impart hybrid properties to the final polymer, being thermally processable like polystyrene but very elastic like butadiene. These are usually referred to as SBS block copolymers denoting their molecular architecture. These thermoplastic elastomers are widely used in utility goods like shoe soles and article of commerce as well as asphalt modification to enhance durability and performance of asphalt roads.

Variants using isoprene also exist as to microstructure variations in which the blocks are somewhat randomized to tailor properties to application requirements.

38.2.4.2 Butyl rubber

Isobutylene (2-methyl propene) can be polymerized using a cationic process initiated with aluminum trichloride in a chlorinated hydrocarbon such as methyl chloride [14]. Polyisobutylene as a homopolymer is used in sealants, caulking, and adhesives and, as it itself contains no unsaturation, cannot be co-vulcanized in tire construction with highly unsaturated elastomers like natural rubber, polybutadiene or SBR. This is overcome by inclusion of a few percent of isoprene in the reaction mixture. The predominant benefit of butyl rubber is the high order of molecular alignment and packing with such uniform distribution of methyl groups in the backbone, as seen in Figure 38.6. This renders butyl rubber and its variants highly air impermeable and suitable for inner liners in tubeless tires to maintain inflation pressure. Halogenation of the butyl rubber in hexane solution makes it even more vulcanization compatible with highly to ensure unsaturated rubbers.

The engineering marvel in production of butyl rubber is that the polymerization is conducted at ca. −100 °C ensuring product quality and performance. At higher temperatures, a chain-transfer side reaction reduces the molecular weight to a region where its utility is primarily as an adhesive and chewing gum base. This temperature is obtained by using liquid ethylene to cool the reactors and it in turn is cooled by liquid ammonia.

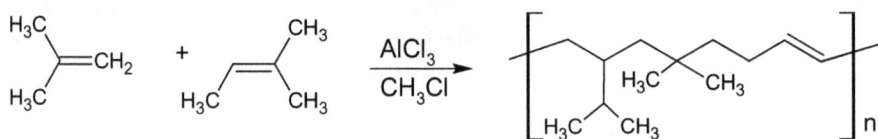

Figure 38.6: Poly(isobutylene co-isoprene), butyl rubber.

38.2.4.3 Polychloroprene (Neoprene® rubber)

Polychloroprene, as shown in Figure 38.7 (Neoprene®), rubber is produced by free radical polymerization using potassium persulfate and is widely used in oil- and solvent-resistant membranes and protective clothing. As it is produced from a 1,3-diene monomer, the resultant polymer is easily cross-linked – in this case with

Figure 38.7: Polychloroprene.

metal oxides like ZnO rather than sulfur. Related in applications and properties is the free-radical chlorination of polyethylene resulting in a rubber-like material used in pond liners, vapor barrier membranes, and solvent-resistant applications usually in conjunction with other elastomers [14]. Chlorosulfonated polyethylene [14] is also classified as a rubber.

38.2.4.4 Fluoroelastomers
Typically prepared in free-radical processes with perfluoro-ethylene, -propylene, or others, these elastomers are generally used in high temperature and aggressive solvent applications.

38.3 Tire construction and compounding

As mentioned, tires are complex composites [4, 14]. In general, the process involves mixing elastomers, fillers (predominantly carbon black), processing aids (lubricants like zinc stearate), sulfur as curative and chemicals to control the curing process (accelerators) in a high intensity mixer such as a Banbury© machine. Up to a ton of material can be mixed in a single batch. This compounded rubber is then masticated further on roll mills to ensure uniformity. This material is then calendered into sheets used in tire construction. Each class of rubber as described above is so compounded to optimize its function in the final tire construction.

Actual tire building, even today involving much hand-layup, entails applying the sheets of rubber onto a building-drum in correct sequence dictated by the final tire, as well as steel and polyamide belting reinforcements and bead wire. The tire "green" tire is then placed in a heated, high-pressure mold for curing, often usually an hour at over 130 °C. The tire is inspected and tested for quality before being shipped.

References

[1]   Koltzenberg, S., Maskos, M., and Nuyken, O. Polmere: Synthese, Eigenschaften und Anwendungen, Springer, Berlin, 2012, 424.
[2]   White, J.L. Intern. Polymer Processing, (14), (1999), 114f.

[3]   Plaumann, H., and Benvenuto, M. "Greening the Tire Industry," 23rd Annual Green Chemistry and Engineering Conference, Reston VA, June 11–13, 2019.

[4]   Kovac, F.J. in, Science and Technology of Rubber, Eirich, F.R. Ed., Am. Chem. Soc., Rubber Division, 1978, p.998.

[5]   The Vanderbilt Rubber Handbook, R. Babitt, ed., 1978, pp. 650–651.

[6]   Tieke, B. Makromolekular Chemie, Auflage, Wiley-VCH, Weinheim, 2014, 149.

[7]   Geng, J., Sun, Y., Hua, J. Polymer Science, Series B, Sep. 2016, 58, 5, 495–502.

[8]   Ullmann's Encyclopedia of Industrial Chemistry, 2011, Wiley-VCH, Weinheim; H-D. Brandt, W. Nentwig, N. Rooney, R.T. LaFlair, U.U. Wolf, J. Duffy, J.E. Puskas, G. Kaszas, M. Drewitt and S. Glander in, "Rubber, 5. Solution Rubbers."

[9]   Yoshioka, A., Komuro, K., Ueda, A., Watanabe, H., Akita, S., Masuda, T., and Nakajima, A. Structure and physical properties of high-vinyl polybutadiene rubbers and their blends. Pure and Applied Chemistry, 58, 12, (1986), 1697–1706.

[10]   Feldman, D., and Barbalata, A. Synthetic Polymers: Technology, properties, applications, Chapman & Hall, London, UK. ISBN: 978–0412710407.

[11]   Kaita, S., Yamanaka, M., Horiuchi, A., and Watasuki, Y. Butadiene polymerization catalyzed by lanthanide metallocene-alkylaluminum complexes with cocatalysts: metal-dependent control of 1,4-Cis/Trans stereoselectivity and molecular weight. Macromolecules, 39, 4, (2006), 1359–1363.

[12]   Ver Strate, G., Cozewith, C., West, R.K., Davis, W.M., and Capone, G.A. Block copolymers of polyethylene and ethylene-propylene-diene elastomer. synthesis, characterization, and properties, Macromolecules, 32, 12, (1999), 3837–3850.

[13]   Guojian, W., Junjie, Y., Wang, G., and Yuan, J. Multicomponent Polymers: Principles, Structures and Properties, DeGruyter and Tongji University Press, 2020, ISBN: 978–3110594416.

[14]   Princi, E. Rubber: Science and Technology, DeGruyter, 2019, ISBN: 978–31106400328.

Chapter 39
Styrene

39.1 Introduction

Styrene is another olefin-based small molecule that is used almost exclusively as a monomer for the production of polystyrene, and that exists as a colorless liquid at room temperature. As with other olefin-based monomers which have some substituent on one carbon atom, styrene can be polymerized into three different tacticities, all based on the arrangement of the phenyl side groups, as shown in Figure 39.1. The only one of commercial importance is atactic polystyrene. The production of polystyrene can be effected using several different catalysts. They are not treated in this chapter. Here, the production of styrene is the focus.

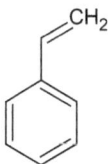

Figure 39.1: Lewls structure of styrene.

39.2 Reaction chemistry

Styrene is produced by the dehydrogenation of ethylbenzene, which is itself produced by the Friedel–Crafts alkylation of benzene. Virtually all the ethylbenzene produced is used to produce styrene, as shown in Figure 39.2.

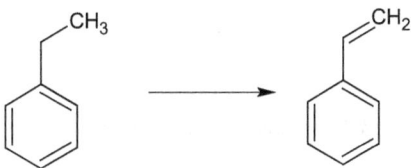

Figure 39.2: Styrene production from ethylbenzene.

This dehydrogenation reaction of ethylbenzene is performed at approximately 600 °C using steam that is superheated, and using an iron (III) oxide catalyst, Fe_2O_3. The reaction is an endothermic one, but can produce styrene at higher than 90% yield. Final purification is achieved through distillation.

https://doi.org/10.1515/9783110542868-039

39.3 Catalyst production

39.3.1 Fe$_2$O$_3$

Iron (III) oxide is a dark red color – a reason it is often called rust in common terms – and goes by several other, older names as well. These include hematite, ochre, and red iron oxide.

Several syntheses for iron (III) oxide exist, with some of them being proprietary. The reason for this is that different companies produce the material for paint pigments, and thus keep careful control over their method, since it produces slightly different shades of a color [1–3]. It is also a feedstock for the reduction of iron and production of iron metal, and the fineness of the particles used in this process is sometimes controlled information [4–7].

One method for the production of iron (III) oxide is shown in Figure 39.3. The final step requires an elevated temperature, ~200 °C.

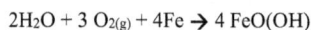

$2H_2O + 3\ O_{2(g)} + 4Fe \rightarrow 4\ FeO(OH)$

$2\ FeO(OH) \rightarrow Fe_2O_3 + H_2O_{(g)}$ **Figure 39.3:** Iron (III) oxide production.

39.4 Catalyst fate

There are other methods for producing styrene, but this is the established one that depends upon a catalyst such as iron (III) oxide.

The use of iron (III) oxide as a catalyst is a minor one when compared to its use in the iron industry. Thus, disposal of it presents no serious problems, since it is inexpensive and readily available.

References

[1]   American Coatings Association. Website. (Accessed 14 September 2020, as: paint.org/about/about-aca/).
[2]   Canadian Coating Associations. Website. (Accessed 14 September 2020, as: coating.ca/Canadian-paint-coating-associations).
[3]   Australian Paint Manufacturer's Federation. Website. (Accessed 14 September 2020, as: apmf.asn.au).
[4]   American Iron and Steel Institute. Website. (Accessed 14 September 2020, as: www.steel.org).
[5]   Canadian Steel Producers Association. Website. (Accessed 14 September 2020, as: www.canadiansteel.ca).
[6]   European Steel Association. Website. (Accessed 14 September 2020, as: www.eurofer.eu).
[7]   Australian Steel Association. Website. (Accessed 14 September 2020, as: www.asa-inc.org.au).

Chapter 40
Sulfuric acid

40.1 Introduction

The large-scale production of sulfuric acid has existed for over 200 years, and in that time, it has seen significant changes and improvements. It has found extensive use in a wide variety of industries, especially in the production of fertilizers. One improvement in production that has occurred is the use of vanadium pentoxide as a catalyst to effect the formation of sulfur trioxide from sulfur dioxide.

40.2 Reaction chemistry

The raw materials for sulfuric acid are elemental sulfur, oxygen, and water. Perhaps obviously, water is considered relatively easy to find (although it must often be purified before use), and oxygen is produced on a large scale from the liquefaction and subsequent distillation of air. Sulfur can be mined, or can be the captured byproduct of crude oil refining.

Traditional mining of sulfur – what is called the Frasch process – involves inserting concentric tubes into an underground sulfur deposit, and blowing superheated water into the deposit, melting the sulfur which is then pushed out of the deposit and dried. This has been the only means by which mineral deposits of elemental sulfur have been captured since the process was pioneered in the 1890s. In the past few decades, more sulfur that is recovered from oil extraction and distillation operations has been used in sulfuric acid production, because allowable emissions of sulfur oxides from refining operations have been more tightly regulated, and emissions to the air curtailed.

The basic reaction chemistry that produces sulfuric acid from these three materials is shown in Figure 40.1.

$$S_{(s)} + O_{2(g)} \rightarrow SO_2$$

$$SO_2 + \tfrac{1}{2} O_{2(g)} \rightarrow SO_3$$

$$SO_3 + H_2O_{(g)} \rightarrow H_2SO_4$$

$$H_2SO_4 + SO_3 \rightarrow H_2S_2O_7$$

$$H_2S_2O_7 + H_2O_{(g)} \rightarrow 2\,H_2SO_4$$ **Figure 40.1:** Sulfuric acid production.

It can be seen that this is a series of addition reactions, as well as oxidation – reduction reactions. The process involves the formation of sulfur trioxide in the second

https://doi.org/10.1515/9783110542868-040

step, and requires a catalyst for this reaction to proceed. That catalyst is vanadium pentoxide (V_2O_5).

40.3 Catalyst production

Vanadium pentoxide, sometimes called vanadium (V) oxide, or vanadia, is a yellow-orange or yellow-brown solid that is manufactured either from some vanadium ore, or from what might be called a vanadium-enhanced residue from a number of other refining processes. The United States Geological Survey Mineral Commodity Summaries [1] notes that the United States is 100% dependent upon imports for vanadium, with Canada, South Korea, the Czech Republic and Austria being the major producers. Vanadium pentoxide is produced by domestic businesses, but is also imported from Russia, South Africa, and China. This import status is also monitored by the United States Department of Defense through their Strategic and Critical Materials 2013 Report on Stockpile Requirements, although in this case it is because vanadium can be alloyed with iron to make ferrovanadium steels, the major use of vanadium worldwide [2].

The Mineral Commodity Summaries also notes that up to 40% of the vanadium in use has been recycled at least once, either from previous vanadium pentoxide, or from any other catalytic use.

The production of usable vanadium pentoxide starts either with an ore, or with a material that contains vanadium as some residue. While it is difficult to display a simple reaction showing the chemistry that isolates the vanadium, it can be shown that sodium carbonate is needed to produce a sodium metavanadate.

$$MV(O, S)_x + Na_2CO_3 \rightarrow NaVO_3$$

The sodium metavanadate must then be precipitated from an acidic solution of $pH \approx 2$, which is made so by the addition of H_2SO_4.

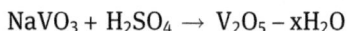

$$NaVO_3 + H_2SO_4 \rightarrow V_2O_5 - xH_2O$$

The final step in V_2O_5 production involves heating the hydrate to 690 °C, resulting in the crude product.

The direct oxidation of vanadium metal with elemental oxygen can also be used to produce vanadium pentoxide. But this process is not used industrially because the oxidation of the vanadium to the +5 oxidation state is seldom complete, meaning vanadium in lower oxidation state oxides must be separated from the product.

40.4 Catalyst fate

The lifetime of the vanadium pentoxide catalyst in the production of sulfuric acid is roughly 20 years. While this certainly speaks to the hardiness of V_2O_5 as a catalyst, it is not an indestructible material. The production of sulfur trioxide is exothermic, and the reaction temperature must be controlled so the catalyst does not break down. Temperatures in excess of 690 °C greatly accelerate the degradation of V_2O_5.

When vanadium pentoxide has reached the end of its useful life, it can be recycled, as mentioned. The important point is returning the vanadium to the +5 oxidation state [3].

References

[1]   U.S. Geological Survey, Mineral Commodity Summaries, 2016, downloadable as: https://minerals. usgs.gov/minerals/pubs/mcs/2016/mcs2016.pdf.
[2]   U.S. Department of Defense Strategic and Critical Materials 2013 Report on Stockpile Requirements, downloadable as: http://mineralsmakelife.org/assets/images/content/ resources/Strategic%20and%20Critical%20Materials%202013%20Report%20on% 20Stockpile%20Requirements.pdf.
[3]   Sulfuric Acid Today, http://issuu.com/sulfuricacidtoday.

Chapter 41
Toluene diisocyanate

41.1 Introduction

Toluene diisocyanate (TDI) is a precursor to polyurethane polymers, and as such, it is an important intermediate in the production of these plastics. More than one isomer of TDI exists, and production methods do not result exclusively in a single isomer as a product. What are called 2,4- and 2,6-isomers are those which are industrially important, are used in large-scale polyurethane production, and are shown in Figure 41.1 [1, 2].

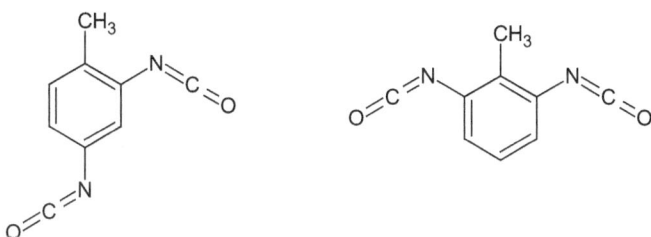

Figure 41.1: Toluene diisocyanate isomers.

In the production of the diisocyanates, other isomers are formed, sometimes in significant quantities, but provide no economic benefit for the producers. Since those other isomers are not economically useful, they are often separated from the above two isomers, then used in some secondary role after separation, such as for a fuel source for further reactions. While this may appear at first glance to be wasteful, no other use has been found for these other isomers, and their use as fuel is economically better than simply disposing of them.

41.2 Reaction chemistry

Toluene is the starting material for the industrially important TDI isomers. Shown in Figure 41.2 is the production of the 2,4-isomer which is the same as for the 2,6-isomer, although all isomers are formed and must be separated.

Of note is that the final step to this material is the addition of phosgene. Because of the toxicity of phosgene, it is often produced and reacted on-site, to eliminate any dangers or accidents that might occur in any sort of transport operations [3]. The production of phosgene has been discussed in Chapter 28.

https://doi.org/10.1515/9783110542868-041

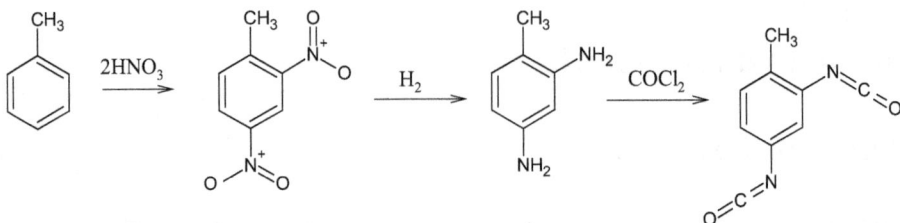

Figure 41.2: Production of 2,4-toluene diisocyanate.

For the purposes of our discussion, the hydrogenation step is important, because it is effected by the use of a nickel catalyst.

41.3 Catalyst production

The catalyst used in numerous hydrogenations is Raney nickel, which has been discussed in earlier chapters, such as Chapter 5. We will reiterate here that it is a nickel–aluminum alloy, a heterogeneous catalyst, and was developed in 1927 by Murray Raney (hence the name).

Additionally, we will make mention that Raney nickel is often produced to have maximum surface area, and that the material is pyrophoric. It must be stored under some inert atmosphere, whether that be liquid, such as mineral oil, or gaseous, such as an inert gas like dry nitrogen or argon.

41.4 Catalyst fate

While nickel can be disposed, since it is not as expensive as several other metals that find use as catalysts, and has low overall toxicity, it has become worthwhile for businesses to see if there is some economically feasible way to re-use or recycle such catalysts [4]. In the recent past, for example, BASF has made serious efforts to recycle such catalysts. Company spokespersons point out the environmental friendliness of such operations. There is an economic advantage to the capture, recycling and reuse of such materials as well, in that this avoids direct disposal costs [5].

References

[1]   American Chemistry Council. Website. (Accessed 13 March 2021 as: plastics.
      americanchemistry.com).
[2]   PlasticsEurope. Website. (Accessed 13 March 2021, as: www.plasticseurope.org).
[3]   International Isocyanate Institute. Website. (Accessed 15 September 2020, as:
      www.diisocyanates.org).
[4]   North American Catalysis Society. Website. (Accessed 15 September 2020, as: nacatsoc.org).
[5]   BASF Catalysts. Website. (Accessed 21 September 2020, as: catalysts.basf.com).

Chapter 42
Vinyl acetate

42.1 Introduction

Vinyl acetate is another small molecule that serves as a monomer for polymer production, often for the single polymer polyvinyl acetate (PVA). The abbreviation is sometimes confused with that for polyvinyl alcohol. The structure for vinyl acetate is shown in Figure 42.1. It is produced at the level of millions of tons annually, with companies in China as well as those in the United States and Europe being contributors to the global market for this commodity. In the United States, Celanese has been the largest producer for several years, while in Europe, LyondellBasell has been the leading producer [1, 2]. Different companies in China have at one time or another been that nation's largest producer.

Essentially all vinyl acetate is used in the production of PVA. Major consumer uses of PVA include the production of latex paints, and as a major component in several adhesives. The well-known Elmer's Glue® is a PVA-containing product. When used in this manner, products are routinely labeled with the phrase "Conforms to ASTM D 4236 Non-toxic," as shown in Figure 42.2. This is the standard used for making possible hazards known in art materials.

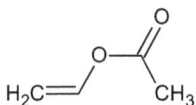

Figure 42.1: Lewis structure of vinyl acetate.

42.2 Reaction chemistry

There have been several different routes for the production of vinyl acetate in the past decades, but one of the major ones remains the addition of ethylene to acetic acid in the presence of oxygen. The basic reaction chemistry is shown in Figure 42.3. Note that water is the by-product in this synthesis.

Also of note, this method requires a palladium catalyst, often one supported on silica. Also, a palladium–gold catalyst supported on silica has proven very effective at producing vinyl acetate, and has been adopted in some cases.

42.3 Catalyst production

The mining and refining of palladium is seldom an operation in which palladium is the primary product, although the United States does have one mine from which

https://doi.org/10.1515/9783110542868-042

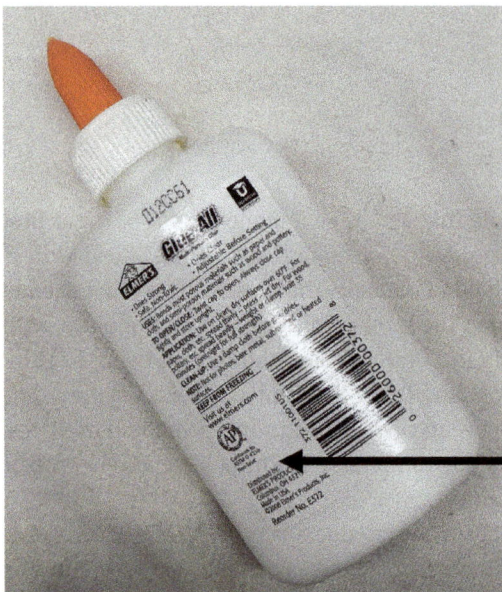

Figure 42.2: Label on an adhesive bottle.

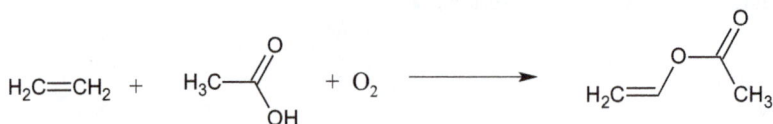

Figure 42.3: Vinyl acetate synthesis.

palladium is the primary element produced. It is one of the platinum group metals often recovered from the anode mud or anode slime of high-purity copper refining, or coproduced with other elements. The element is valuable enough that the United States Geologic Survey tracks it in its annual *Mineral Commodity Summaries* [3]. In the past decade, a few Mints from different governments, including those of Canada and the United States, have produced palladium coins as a form of investment in precious metals. Companies such as Johnson Matthey, which produce numerous different types of catalysts, also sell palladium ingots and bars to investors and collectors.

Gold is mined extensively in the United States and elsewhere, and is also recovered as a by-product in copper refining, in the anode mud. Gold use and value is tracked extensively, and trade organizations exist to promote its uses [3–5].

42.3.1 Catalyst form

As a catalyst for vinyl acetate production, palladium is impregnated onto silica, often as some salt, which is then reduced to the metal. This ensures that the surface area of both palladium and palladium–gold on the silica is maximized. The reason for this is the obvious one, that the effectiveness of the catalyst should be made as high as possible. The slightly less obvious reason is that the cost of these metals is high enough that it is economically feasible to use as little of them as possible.

Research has been done extensively to determine what crystal morphologies and deposition techniques result in the greatest catalytic activity [6–13], but some information remains proprietary to companies. What is publically available always indicates that palladium or a palladium–gold mix must have as great a surface area as possible.

42.4 Catalyst fate

As with many of the other processes we have examined, the cost of the palladium catalyst is high enough that material recovery is important. Likewise, gold is an expensive enough material that both are recovered when their catalytic activity is spent. Patents addressing the improvement of vinyl acetate production state this directly: "Palladium and gold are expensive precious metals. Therefore, many efforts have been made to increase the catalytic activity and reduce the amount of catalyst needed" [12].

For decades, the cost of gold was far higher than the cost of palladium, but in the years 2019 and 2020, the cost of the two metals has essentially reversed themselves. At the end of 2020, gold had seen a price spike of over $2,000 per troy ounce, and had come back down to approximately $1,700 per troy ounce [4]. At the same time, palladium had risen to over $2,200 per troy ounce and remained there. Thus, both metals are always recovered and re-refined if necessary.

References

[1]   Celanese Corporation. Website. (Accessed 14 December 2020, as: celanese.com).
[2]   LyondellBasell. Website. (Accessed 14 December 2020, as: lyondellbasell.com).
[3]   U.S. Geological Survey. Mineral Commodity Summaries, 2020, downloadable at: pubs.er. usgs.gov, as https://doi.org/10.3133/mcs2020.
[4]   World Gold Council. Website. (Accessed 13 December 2020, as: www.gold.org).
[5]   Gold Prospectors Association of America. Website. (Accessed 13 December 2020, as: https:// www.goldprospectors.org).
[6]   Han, Y.F., Kumar, D., Sivadinarayana, C., and Goodman, D.W. Kinetics of ethylene combustion in the synthesis of vinyl acetate over a Pd/SiO_2 catalyst. Journal of Catalysis, 224, (2004), 60–68.

[7]   U.S. Patent. Sennewald, K., Vogt, W., and Glaser, H. Palladium-gold catalyst. US3743607, 1973.

[8]   U.S. Patent. Fernholz, H., Wunder, F., and Schmidt, H-J. Oxacylation of olefins in the gaseous phase. US3939199, 1976.

[9]   U.S. Patent. Nicolau, I., Colling, P.M., and Johnson, L.R. Palladium-gold catalyst for vinyl acetate production. US5693586, 1997.

[10]   U.S. Patent. Zeyss, S., Dingerdissen, U., and Fritsch, J. Process for the production of vinyl acetate. US6852877B1, 2005.

[11]   European Patent Application. Method for producing vinyl acetate. EP2778153A1, 2012.

[12]   U.S. Patent. Salisbury, B.A., Hallinan, N.C., and Oran Osment, J.M. Vinyl acetate production process. US8822717B2, 2014.

[13]   U.S. Patent. Dellamorte, J.C., and Mentz, R.T. Method and catalyst composite for production of vinyl acetate monomer. US2014/0039218A1, 2014.

Chapter 43
Vinyl chloride monomer

43.1 Introduction

Vinyl chloride monomer (often abbreviated VCM) is also called vinyl chloride or chloroethene, and has the formula CH_2CHCl. It is another small olefin, much like those discussed in other previous chapters, and is almost exclusively used for the production of polyvinyl chloride (PVC). PVC is one of the largest commodity plastics produced worldwide, with numerous applications in commercial and consumer items, an example of which – piping – is shown in Figure 43.1. Over 10 billion kilograms are routinely produced per annum. The simplified reaction chemistry for this synthesis is shown in Figure 43.2. In this figure, we do not show the tacticity of the PVC product, since most PVC is atactic with small domains of syndiotactic repeat units, which impart some crystallinity to the finished material.

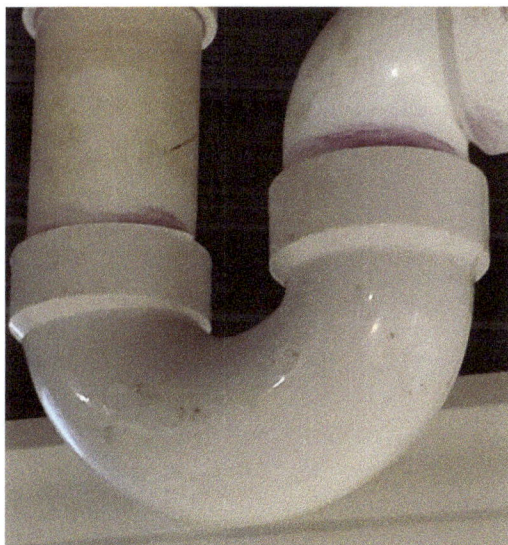

Figure 43.1: PVC piping in a residence.

Because of the wide suite of uses for PVC, the production of VCM is itself enormous. Once again, billions of kilograms are produced annually. Also, because of the wide-spread need for PVC, there are many different manufacturers for it. Currently, slightly under half of the world's production of VCM comes from China [1]. The United States is the second largest consumer of VCM, again for the production of PVC [1]. Yet the

https://doi.org/10.1515/9783110542868-043

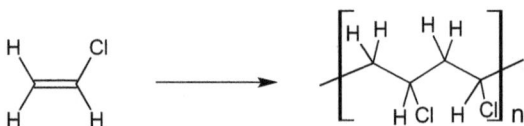

Figure 43.2: Polyvinyl chloride synthesis.

overall production of VCM is widespread enough that there are organizations devoted to its production and use, as well as that of PVC [2, 3].

43.2 Reaction chemistry

The reaction chemistry that produces VCM may use various starting materials, but can be represented both by the dehydrochlorination of 1,2-dichloroethane, and by the hydrochlorination of acetylene. The second requires a mercury (II) chloride catalyst, and is a well-established reaction [4]. Both reactions are shown in Figure 43.3. Note that while the second requires the catalyst, it does not produce hydrochloric acid as a by-product. Also, the mercury (II) chloride is routinely supported on activated carbon.

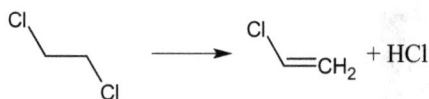

Figure 43.3: Methods of VCM production.

Because of the high toxicity of mercury (II) chloride, the search continues for catalysts that are less toxic than a mercury (II) salt, and progress toward this end has been made in the recent past [5–7]. However, China has large reserves of coal, from which acetylene can be made easily. This means that there is an economic incentive to continue to use the process which requires this mercury catalyst.

Interestingly, the original production of VCM actually goes far back to an old synthesis in the early nineteenth century, pioneered by Justus von Liebig. This has been superseded by the two different synthetic routes that are in use today.

43.3 Catalyst production

43.3.1 Mercury (II) chloride, HgCl$_2$

Despite its toxicity, this mercury compound is used, perhaps obviously, because it is very effective in the production of VCM. It can be prepared by direct addition of the elements, as follows:

$$Hg_{(l)} + Cl_{2(g)} \longrightarrow HgCl_2$$

Mercury (II) chloride can also be prepared via the addition of a solution of hydrochloric acid at elevated temperature to some mercury (I) salt. The nitrate works well, as shown below:

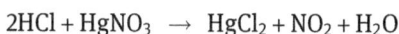

$$2HCl + HgNO_3 \longrightarrow HgCl_2 + NO_2 + H_2O$$

Other methods exist as well, but these two tend to be the major routes to this salt. In all cases, care must be taken to ensure worker safety during the handling of mercury, especially in its oxidized +2 state where even skin exposure can have catastrophic results.

43.4 Catalyst fate

Mercury must be recovered after use, because it can cause significant environmental damage if spilled or dumped, or even if stored in hazmat-designated landfills. Thus, when VCM is produced using mercuric chloride, the catalyst is recovered. The use of this has slowly ceased in the United States and Europe, but as mentioned remains a major method by which VCM is manufactured in China. The Minamata Treaty – more properly the Minamata Convention on Mercury – was approved in 2013, and specifies use and control of mercury in industrial applications. Concerning VCM, it states, "VCM production using the acetylene process involves the use of mercuric chloride as a catalyst. The Minamata Convention requires parties to take measures to reduce the use of mercury in terms of per unit production by 50 per cent by the year 2020 against 2010 use" [8]. It also suggests the prevention of any escape of mercury through a variety of means, including for example capture on activated carbon.

References

[1] IHS Markit. Vinyl Chloride Monomer (VCM). Website, (accessed 3 July 2020, as ihsmarkit.com).

[2] Vinyl Institute. Website. (Accessed 19 December 2019, as: https://www.vinylinfo.org/resour ces/pvc-pipe/).

[3] Eurochlor. Website. (Accessed 19 December 2019, as: http://www.eurochlor.org/).

[4] U.S. Patent. Weller, J.F. Production of vinyl chloride. US2412308, 1946.

[5] Chemistry World. New vinyl catalyst will reduce mercury emissions. Website. (Accessed 19 December 2019, as: https://www.chemistryworld.com/news/new-vinyl-catalyst-will-reduce-mercury-emissions/9326.article).

[6] Zhang, J., Liu, N., Wei, L., and Dai, B. Progress on cleaner production of vinyl chloride monomers over non-mercury catalysts. Frontiers of Chemical Science and Engineering, 5, (2011), 514–520. https://link.springer.com/article/10.1007/s11705-011-1114–z

[7] Brett Graham, L., and Hutchings, J. Vinyl chloride monomer production catalyzed by gold: a review. Chinese Journal of Catalysis, 37, 10, (2016) 1600–1607. DOI: 10.1016/S1872-2067(16) 62482-8.

[8] UN Environmental Programme, Minamata Convention on Mercury. Website (accessed 3 July 2020, as: mercuryconvention.org).

Chapter 44
Vitamins

44.1 Introduction

Throughout history, people have had a vague understanding of what have since become known as vitamins, but routinely only through trial and error. Perhaps the
most famous historical example is the plague of scurvy that for hundreds of years
incapacitated and sometimes killed sailors on long voyages. It is caused by a lack of
vitamin C. Since this vitamin was an unknown substance at the time, various reme-
dies for scurvy were tested. British naval surgeon James Lind found through trial
and error that eating citrus fruits or drinking their juice prevented scurvy. This
solved the problem – and incidentally introduced the nickname "limeys" into the
vocabulary, as a name for men in the British navy. Curiously though, it was the
Spanish navy that first found this solution to scurvy in the late 1700s.

The term "vitamin" comes from the belief that all such substances were "vital
amines." The understanding of these essential nutrients has expanded beyond just
amines as time has gone on. Today, several of them remain complex enough in
their syntheses that some catalyst is required for a step, often a biocatalyst.

44.2 Reaction chemistry and syntheses

44.2.1 Vitamin A

Vitamin A is actually a group of organic molecules, all of which contain degrees of
unsaturation, and which include both retinol and retinal, as well as retinoic acid.
Figure 44.1 shows the Lewis structure.

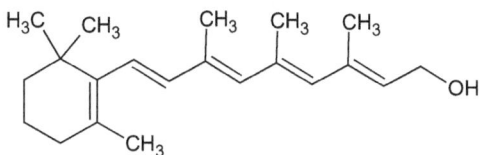

Figure 44.1: Vitamin A, retinol form.

The synthesis of vitamin A dates back to 1947, and the large-scale synthesis of vitamin A was first undertaken by Hoffmann-LaRoche. The synthesis continues to evolve
today [1, 2].

https://doi.org/10.1515/9783110542868-044

In the synthesis of vitamin A, one step requires the use of Lindlar's catalyst. This is palladium (Pd) on calcium carbonate ($CaCO_3$), but poisoned with trace amounts of lead. The catalyst is broadly useful for the reduction of alkynes to alkenes in a stereoselective manner, giving exclusively the *cis*-form of the resultant double bond.

Production of Lindlar's catalyst is difficult to show as stoichiometrically balanced equations, but is essentially as follows:

1. Addition of palladium (II) chloride to calcium carbonate, in a slurry.
2. Addition of a small amount of a lead salt, usually $Pb(C_2H_3O_2)_2$, although other lead salts can fulfill the role.
3. Ultimately, palladium is no more than 5% of the entire catalyst by mass.

44.2.2 Vitamin B₁ or thiamine

The complexity of vitamin B_1 requires any variety of fungi or other biocatalysts for its large-scale production. Purely synthetic chemical methods are too expensive for such a synthesis. Hoffman-La Roche has pursued this synthesis successfully, filing patents for it. The Lewis structure for vitamin B_1 is shown in Figure 44.2.

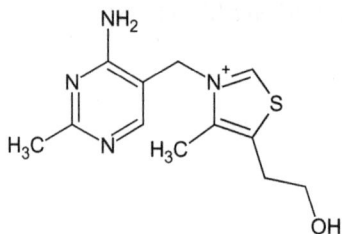

OH   **Figure 44.2:** Vitamin B_1.

The ring structure of vitamin B_1 is part of what makes it a difficult molecule to produce via purely synthetic, organic transformations. Current patents use microorganisms that hyper-express thiamine, and routinely work via fermentation [3].

44.2.3 Vitamin B₂ or riboflavin

Another of the B vitamins, vitamin B_2 is also produced industrially. The following microorganisms have been used to produce it successfully:

Ashbya gossypii – aka *Eremothecium gossypii*, relative of yeast. Utilized by BASF.
Bacillus subtilis – aka grass bacillus or hay bacillus
Candida famata – type of yeast
Candida flaveri
Corynebacterium ammoniagenes

The Lewis structure for riboflavin is shown in Figure 44.3.

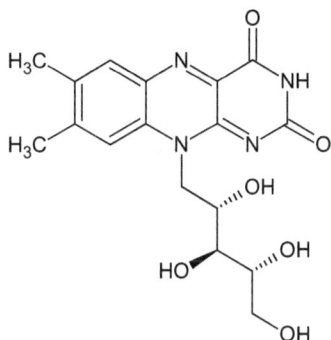

Figure 44.3: Riboflavin.

A large amount of the vitamin B_2 produced each year is for animal feed, and not for direct human consumption.

44.2.4 Vitamin B_3, often niacin

Vitamin B_3 exists in three forms: niacin, nicotinamide, and nicotinamide riboside. Niacin, also known as nicotinic acid, can be produced in pure enough form that it is marketed as a prescription drug for people who require it. The Lewis structure of niacin is shown in Figure 44.4.

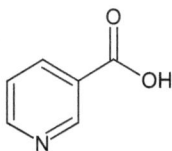

Figure 44.4: Niacin.

In the industrial-scale production of niacin, nitrile hydratases are used in the multi-step synthesis, often one that begins with tryptophan. These enzymes catalyze the transformation of nitriles to amines.

44.2.5 Vitamin B_5 or pantothenic acid

Several different enzymes are used in the production of vitamin B_5 when a biosynthetic pathway for its production is utilized. The synthesis requires several steps, and thus they include:

Dihydroxy-acid dehydratase – loss of –OH group
Valine–pyruvate transaminase – replacement of amine by ketone
Hydroxymethyltransferase – addition of hydroxyl methyl group
Ketopantoate reductase – selective ketone to hydroxyl reduction
Pantothenate synthase

These biocatalysts are required for the needed, multiple transformations from 1,2-dihydroxyvalerate starting material to pantothenic acid.

As a calcium salt, calcium pantothenate, vitamin B_5 has been approved for use as an animal feed, by the European Food Safety Authority. The Lewis structure of vitamin B_5 is shown in Figure 44.5.

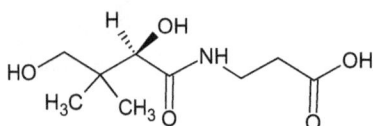

Figure 44.5: Pantothenic acid.

44.2.6 Vitamin B_6 or pyridoxine

The synthesis of vitamin B_6 goes at least as far back as 1939, when it was reported by Harris and Folkers, then working for Merck & Co. [4]. Despite it being found widely in a number of food sources, its synthesis is still pursued on a large scale. What has been called the Kondrat'eva approach has been used in some way since the 1950s [5, 6] – Diels–Alder followed by aromatization of the ring.

The Lewis structure of vitamin B_6 is shown in Figure 44.6.

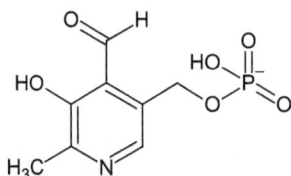

Figure 44.6: Lewis structure of pyridoxine.

44.2.7 Vitamin B_7 or biotin

Curiously, vitamin B_7 sometimes still goes by the older name, "vitamin H" because of its positive effects on hair and skin (Haar und Haut, in German). Understanding of its role in health dates back only to the 1940s. The structure of vitamin B_7 is shown in Figure 44.7.

Biosynthetically, vitamin B_7 can be produced from pimeloyl-CoA and alanine, and requires the following to do so:

- 8-Amino-7-oxopelargonic acid synthase
- 7,8-Diaminopelargonic acid aminotransferase
- Dethiobiotin synthetase

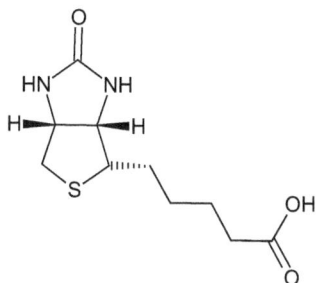

Figure 44.7: Lewis structure of vitamin B_7.

The odd, fused ring structure has traditionally been one of the biggest challenges in the synthesis of this molecule.

44.2.8 Vitamin B_9 or folic acid

Once again, although folic acid – or folacin – is found in a wide variety of foods, since it is an essential vitamin, it has been manufactured on a large scale. Figure 44.8 shows the Lewis structure.

Once again, the synthesis of this vitamin is through several steps, some utilizing biocatalysts. Starting materials are 2-amino-4-hydroxy-6-(chloromethyl)-pteridine and p-aminobenzoyl-l-glutamic acid [7].

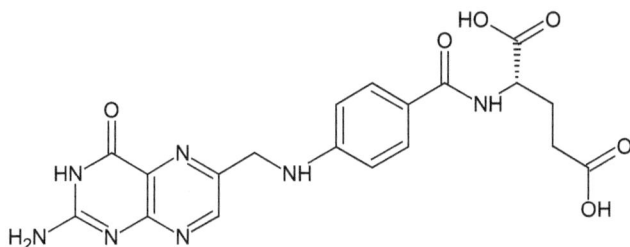

Figure 44.8: Lewis structure of vitamin B_9.

44.2.9 Vitamin B_{12} or cyanocobalamin

Undoubtedly, the structure of vitamin B_{12}, also known as cobalamin, is the most complex of any of the vitamins. The original, total chemical synthesis of it involved hundreds of steps, and was completely impractical for making it on a large scale. The intense study of it and its structure has also been the basis of not one but five Nobel prizes. Today, it is produced via fermentation of several different microorganisms, most of which have been altered to increase expression of the vitamin. They include:

- *Propionibacterium freudenreichii* (shermanii) – also used in production of cheeses
- *Pseudomonas denitrificans* – used by Rhone-Poulenc. The bacterium overproduces vitamin B_{12} for use in synthesis of methionine
- *Streptomyces griseus* – one of the early bacteria used for large-scale production

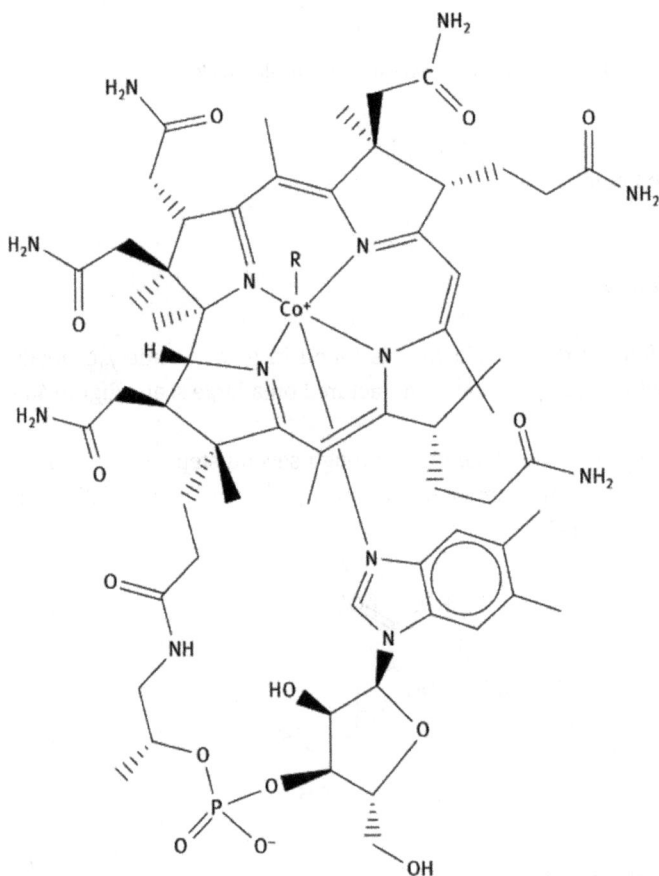

Figure 44.9: Lewis structure of vitamin B_{12}.

Clearly, as seen by the complexity of the structure in Figure 44.9, there is no artificial method to produce this vitamin that can be economically competitive with the biosynthetic pathways of the organisms, above.

44.2.10 Vitamin C or ascorbic acid

The story of vitamin C's discovery and use is a famous one, since the lack of vitamin C in the human diet causes scurvy, and has done so for centuries. This plague of sailors throughout history was only halted by the efforts of a few men who qualify as both medical doctor and scientist, the most famous of them is British naval surgeon, Dr. James Lind, whose work in the mid-1700s is considered one of the first ever clinical, medical trials. Less well known is Georg Wilhelm Steller, who sailed with Vitus Bering at roughly the same period of time, and who also initiated what might be called primitive trials to determine the cause of scurvy in Bering's crew.

The industrial-scale production of vitamin C has a long history, with the Reichstein–Gruessner process being the first to produce it. This uses *Gluconobacter oxydans* in the production, effectively a biocatalyst.

More recently, the fermentation of *Ketogulonicigenium vulgare* and *Bacillus* has been used to convert L-sorbose to 2-keto-L-gulonic acid (often abbreviated 2-KLG), in what is a two-step process. Again, the biological component acts in a catalytic manner. Figure 44.10 shows the transformation.

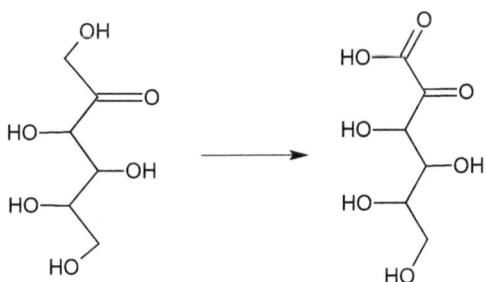

Figure 44.10: Vitamin C production, biocatalytic conversion.

As with the other vitamins, vitamin C is a complex enough molecule that its production from simple starting materials involves several steps. Note the structure of vitamin C, in Figure 44.11, is significantly different from the intermediate product in Figure 44.10.

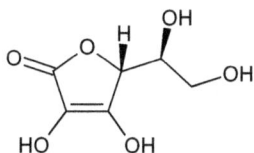

Figure 44.11: Lewis structure of vitamin C.

44.2.11 Vitamin D or cholecalciferol

What is called vitamin D is actually a group of steroidal compounds known as se-costeroids – individually: cholecalciferol (vitamin D_3), ergocalciferol, vitamin D_4, and vitamin D_5. It is well known that the natural production of vitamin D in humans involves exposure to sunlight, specifically the UV-B radiation in sunlight [8].

Industrially, vitamin D_3 is produced by UV-B irradiation of 7-dehydrocholesterol, which itself can be obtained from the lanolin in sheep's wool. This is not a catalyst in the traditional definition of the word, but does function in a similar way in that it is not taken up in the chemical product.

The Lewis structure of vitamin D_3 is shown in Figure 44.12.

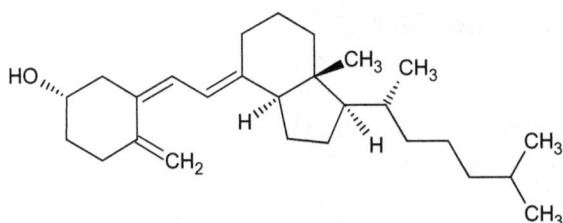

Figure 44.12: Lewis structure of vitamin D_3.

Notice in the Lewis structure of vitamin D_3 that the A-B-C-D rings of a typical steroid are not all formed.

44.2.12 Vitamin E or tocopherols

Vitamin E is actually a group of eight related compounds. Four are classified as to-copherols, and the other four are classified as tocotrienols, the major difference in their structures being the unsaturation in the side chain. The tocotrienols have three sites of unsaturation. Figure 44.13 shows the Lewis structure of one of the to-copherols (the R,R,R isomer).

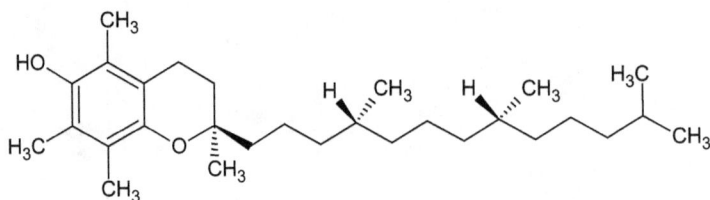

Figure 44.13: Lewis structure of vitamin E.

In a non-biochemical synthesis, toluene can be used as a starting material. As well, 2,3,5-trimethyl hydroquinone is a necessary starting material. In the presence of iron as a catalyst coupled with HCl in the gas phase, the product can be obtained in purity greater than 95%.

Both iron and hydrogen chloride are inexpensive and easy to produce.

44.2.13 Vitamin K or phylloquinone

Vitamin K was found to be related to blood coagulation, and thus takes its letter-name from the Danish word "koagulation." It is a group of compounds, all fat soluble, all with a long side chain attached to a fused ring structure. Differences between one type of vitamin K and another tend to be in the structure of the side chain. Figure 44.14 shows the structure for vitamin K_1, sometimes known as phytomenadione.

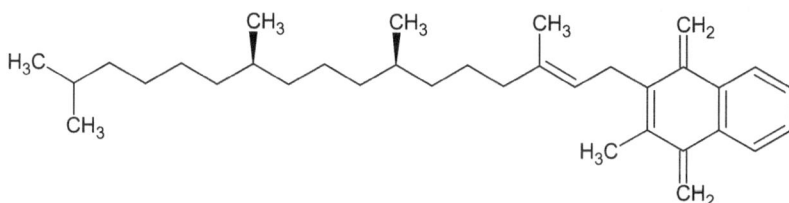

Figure 44.14: Lewis structure of vitamin K_1.

The conversion of vitamin K types can be affected by the gut bacteria, which function enzymatically, and thus catalytically. The improvements for the production of the family of vitamin K molecules are ongoing, and patents continue to be filed in this area [9].

44.3 Catalyst production

One aspect of almost all the catalysts mentioned here that they have in common is that they are bio-based, either being enzymes or microorganisms. As such, production of such catalysts is often a matter of purifying an enzyme or culturing some microorganism so that it can continue to express the desired target molecule.

References

[1] Council for Responsible Nutrition. Website. (Accessed 19 December 2019, as: https://www.crnusa.org/).

[2] Natural Products Association. Website. (Accessed 5 March 2020, as: https://www.npanational.org/).

[3] Thiamine production by fermentation. Patent. EP1651780B1, 2010 (accessed 3 July 2020).

[4] Harris, S.A., and Folkers, K. Synthesis of vitamin B_6. Journal of the American Chemical Society, 61, 5, (1939), 1245–1247.

[5] Kondrat'eva, G.Y. Khim. Nauk Prom., 2, (1957), 666.

[6] Kondrat'eva, G.Y. Izv. Akad. Nauk SSSR, Ser. Khim., (1959), 484.

[7] Spiegelberg, Hans. Process for the manufacture of folic acid. US Patent 2,487,393A, 1949.

[8] The Vitamin D Society. Website. (Accessed 5 March 2020, as: https://www.vitamindsociety.org/about_us.php).

[9] Dotz, Karl H. Process for Preparing Vitamin K. US Patent 4,374,775, 1981.

[10] American Heart Association. Website. (Accessed 5 March 2020, as: https://www.heart.org/en/healthy-living/healthy-eating/eat-smart/nutrition-basics/vitamin-supplements-hype-or-help-for-healthy-eating).

Index

https://doi.org/10.1515/9783110542868-045

www.ingramcontent.com/pod-product-compliance
Lightning Source LLC
Chambersburg PA
CBHW061418210326
41598CB00035B/6254